Weed Seeds
of the Great Plains

Weed Seeds of the Great Plains

A Handbook for Identification

Linda W. Davis

Published for the Cooperative Extension Service
of Kansas State University by the
University Press of Kansas

Agricultural Experiment Station contribution number 92–125–B

Published by the University Press of Kansas (Lawrence, Kansas 66049), which was organized by the Kansas Board of Regents and is operated and funded by Emporia State University, Fort Hays State University, Kansas State University, Pittsburg State University, the University of Kansas, and Wichita State University

Library of Congress Cataloging-in-Publication Data

Davis, Linda W.
    Weed seeds of the Great Plains: a handbook for identification /
    Linda W. Davis
        p.   cm.
    Includes bibliographical references and index.
    ISBN 0-7006-0651-3 (alk. paper)
    1. Weeds—Great Plains—Seeds—Identification.   I. Title.
SB612.G73D38   1993
632'.58'0978—dc20   98-15615

British Library Cataloguing in Publication Data is available.

Printed in the United States of America
10 9 8 7 6 5 4 3 2 1

Contents

Acknowledgments

T. M. Barkley provided much helpful advice in the preparation of this book, from the time the work began as a master's thesis in the Division of Biology at Kansas State University to its completion.

Among the many people who offered suggestions and shared their expertise are Lowell Burchett of the Kansas Crop Improvement Association, Jack Brotemarkle, Eileen K. Schofield, Francis Barnett, and Ralph Brooks.

The line drawings in the Illustrated Glossary were prepared by Iralee Barnard.

Acknowledgments are also due to Robert B. Kaul, Duane Isely, and C. R. Gunn for reviewing and commenting on the manuscript.

The Cooperative Extension Service has provided financial support for publication, specifically for printing the color photographs, without which the book would be much less useful. Dean Walter Woods, director of the Cooperative Extension Service at Kansas State University, was helpful in arranging for this support. Special acknowledgments are also due to George Ham, associate director of the Agricultural Experiment Station, and to Terry Johnson, director of the Division of Biology at Kansas State University, for their assistance.

Introduction

The Great Plains is a large region of natural grasslands. Over the past century and a half, European settlement has transformed much of the region into crop fields and rangeland. As the settlers plowed the prairies, planted grains, and introduced livestock, they also introduced new plants, both crops and weeds.

Agricultural activities are invariably accompanied by weeds, and the maintenance of productivity requires constant attention to weed control. A weed may be defined as any plant that interferes with agriculture or other human enterprises, that is considered undesirable, or that spreads rapidly and aggressively. Weeds may decrease crop yields by competing with the crop for water and nutrients or by impeding planting or harvest. In pastures and rangeland, weeds take the place of nutritious plants and may be toxic to livestock. Weeds may also infest gardens, lawns, and roadsides.

Correct identification of weeds is important to their effective and economical control, whether that control is by means of mechanical removal, land management practices, or herbicides. Herbicides in particular must be targeted to specific weeds, as most herbicides are selective in their action, are potentially hazardous, and require a major monetary investment. The best methods of weed control are those that prevent weeds from becoming established in the field. Identification of weed seed contaminants of crop seed is important in this regard, and in fact one of the most effective weed control measures is the inspection and certification of crop seeds performed by governmental agencies and commercial laboratories.

Weed seed identification is possible because seeds of different species, even closely related species, are remarkably distinctive. There is also marked uniformity of appearance among seeds of a given species; seed characters are more consistent than such plant characters as size and shape of leaves. Thus seeds of different species

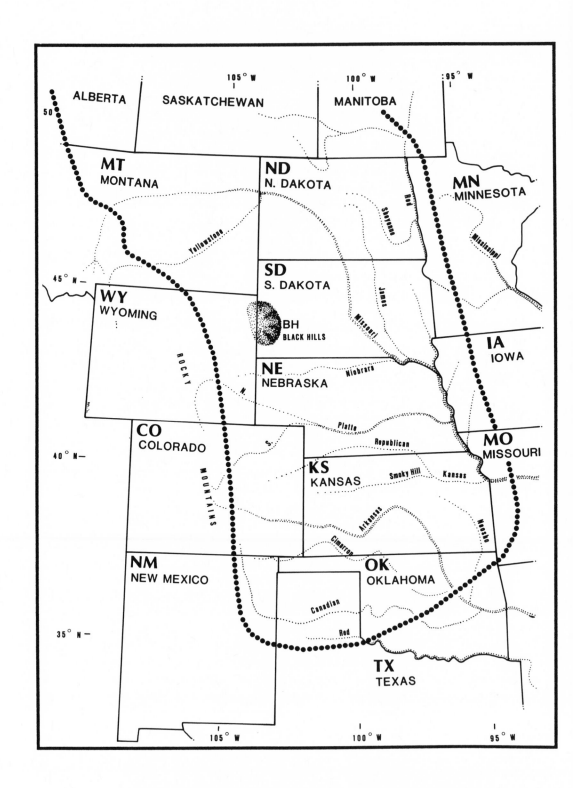

Figure 1. Geographical area covered in *Weed Seeds of the Great Plains.*

can nearly always be distinguished from one another, if carefully examined.

The purpose of this book is to provide detailed information about the seeds of weedy plants of the Great Plains and to enable the reader to compare, distinguish, and identify seeds of interest. The Great Plains is defined here as it was in *Flora of the Great Plains* (Great Plains Flora Association, 1986, 1992). Figure 1 shows the approximate boundaries of the region. Seeds of 280 weedy plants of the Great Plains are included. All are weeds of cropland, range, lawn, garden, or roadside that are abundant or widespread in the region. Some less common plants are included because they are particularly problematic; certain other plants are included because they are conspicuous and common, though not especially troublesome. Aquatic plants, trees, and most shrubs are excluded.

For each of these 280 species, this book provides a written description of the external appearance of the seed, an actual-size silhouette photograph, and a color photograph at low magnification. Finding lists facilitate identification of an unknown seed; a glossary of technical terms and an illustrated glossary of descriptive terms for shapes and textures are also included.

Methods Seed samples along with voucher plant specimens were collected by the author for about 190 of the species. The identity of the plants was verified in the Herbarium of Kansas State University. For the remaining species, seed collections of other botanists, notably H. A. Stephens, were obtained. Most of the seed collections and corresponding voucher plants are in the permanent collection of the Herbarium of Kansas State University (hb KSC). The remaining seed samples and voucher specimens are in the collection of the Herbarium of the University of Kansas (hb KANU). In all cases, the seeds were collected from naturally occurring weedy populations in the region; in nearly all cases, the samples consist of at least 10 mature seeds. For some species, several seed samples were obtained and examined. For a number of species, the natural unit of dispersal is a fruit rather than a seed. For these species, the fruits were collected, examined, and described in the same manner as for seeds.

All seeds were examined at several magnifications ranging from 1× to 25×. The descriptions are based mainly on features visible at 1× to 10× for larger seeds, and at 4× to 10× for smaller seeds. A binocular microscope with an ocular micrometer was used for the observations and measurements. At least 10 seeds were examined and measured for each species. A range of measurements was obtained, as the samples were not large enough to permit the calculation of averages.

For each species, an actual-size silhouette photograph was made. These silhouettes were prepared by placing one to several seeds on glass suspended over a brightly lit white background. The camera was a 35-mm single lens reflex with a 100-mm lens; the film was Kodak Technical Pan 2415.

For each species, a color photograph was made. The photographs were made at 1×, 2.4×, or 4.8× magnification, depending on the size of the seeds. The photographs at 1× magnification were made with a 35-mm single lens reflex camera with a 100-mm lens. The photographs at 2.4× and 4.8× magnifications were made with a Wild dissecting microscope with built-in camera. In all cases, the film was Kodak Ektachrome Tungsten 160, and the light source was a Volpi HL 100 fiber optics unit. Enlargement of these photographs in printing resulted in final magnifications of approximately 2×, 4.8×, and 9.6×.

The taxonomy and nomenclature of *Flora of the Great Plains* were followed throughout. The sequence of the families, as in *Flora of the Great Plains*, was adopted from Cronquist (1981). The common names were drawn from those in *Flora of the Great Plains* and other names that are in use in the region.

References

Beal, W. J. 1910. *Seeds of Michigan Weeds*. East Lansing: Michigan Agricultural Experiment Station Bulletin 260.
Descriptions and drawings of many species.

Esau, K. 1977. *Anatomy of Seed Plants*. New York: John Wiley & Sons.
The standard reference for plant anatomy.

Fahn, A., and E. Werker. 1972. Anatomical mechanisms of seed dispersal. In *Seed Biology*, vol. 1: 151–221, edited by T. T. Kozlowski. New York: Academic Press.
Discussion of seeds and fruits as units of dispersal.

Cronquist, A. 1981. *An Integrated System of Classification of Flowering Plants*. New York: Columbia University Press.
A modern system of classifying and ordering the families and genera of flowering plants. It is the system used in this book.

Delorit, R. J. 1970. *An Illustrated Taxonomy Manual of Weed Seeds*. River Falls, Wisc.: Agronomy Publications.
Color photographs, descriptions, and keys for 192 species. The geographic area covered is mainly the northeast and north central United States.

Delorit, R. J., and C. R. Gunn. 1986. *Seeds of Continental United States Legumes (Fabaceae)*. River Falls, Wisc.: Agronomy Publications.
Color photographs and concise descriptions of the seeds of 206 species of legumes of the United States.

General Purpose Multiple-Entry Key Algorithm and Editor (MEKA 1.2, MEKAEDIT 1.0) Berkeley, Calif.: Thomas Duncan and Christopher Meacham.
This key-writing algorithm was used to help construct the Finding Lists for this book.

Great Plains Flora Association. 1977. *Atlas of the Flora of the Great Plains*. Ames: Iowa State University Press.
Distribution maps for native and naturalized plants in the region.

_____ . 1986, 1992. *Flora of the Great Plains*. Lawrence: University Press of Kansas.
Complete descriptions of all the flowering plants of the region.

Gunn, C. R. 1972. Seed collection and identification. In *Seed Biology*, vol. 3: 55–143, edited by T. T. Kozlowski. New York: Academic Press.
A summary of standard scientific techniques for the study of seeds.

Hitchcock, A. S., and G. L. Clothier. 1897. Kansas Weeds, IV— *Fruits and Seeds*. Manhattan: Kansas Agricultural Experiment Station Bulletin 66.
The earliest regional work on weed seeds.

Korsmo, Emil. 1935. *Ugressfro* [Weed Seeds]. Oslo: Grondahl & Sons.
Extensive coverage of European weed seeds, some of which are also seen in the Great Plains. Excellent descriptions and color plates. Text in English, Norwegian, and German in parallel columns.

Kozlowski, T. T., and C. R. Gunn. 1972. Importance and characteristics of seeds. In *Seed Biology*, vol. 3: 1–20, edited by T. T. Kozlowski. New York: Academic Press.
Excellent introduction to the study of seeds.

Martin, A. C., and W. D. Barkley. 1961. *Seed Identification Manual*. Berkeley: University of California Press.
Black-and-white photographs and life-size silhouettes of 600 species; also a section devoted to family characteristics and useful clues for identification. Geographic coverage is the continental United States.

Montgomery, F. H. 1977. *Seeds and Fruits of Plants of Eastern Canada and Northeastern United States*. Toronto: University of Toronto Press.

Black-and-white photographs of the seeds of a large number of species. Brief descriptions and measurements for each species.

Murely, M. R. 1951. Seeds of the Cruciferae of Northeastern North America. *American Midland Naturalist* 46: 1–81.
Detailed descriptions of Cruciferae seeds.

Nebraska Department of Agriculture. 1975. *Nebraska Weeds*. Lincoln: Nebraska Department of Agriculture.
Color photographs of many seeds along with photographs and descriptions of weedy plants.

University of Illinois at Urbana-Champaign. College of Agriculture. Agricultural Experiment Station. 1981. *Weeds of the North Central States*. Urbana: North Central Regional Research Publication 281, Bulletin 772.
Detailed line drawings of many seeds along with drawings and descriptions of weedy plants.

Radford, A. E., W. C. Dickson, J. R. Massey, and C. R. Bell. 1974. *Vascular Plant Systematics*. New York: Harper & Row.
A useful reference for descriptive terminology and definitions.

Finding Lists

To use these finding lists to identify an unknown seed, first examine the seed and observe its three-dimensional form. Take note of any unusual structures and of the surface texture, size, and color. From the 22 descriptive phrases in the list below, find one that accurately describes the unknown seed. This will give the number of a group to start with; then turn to the subgroups on the following pages to narrow the choices. For example, if the most noticeable feature of the unknown seed is that it is spherical, start with group 7. Check subgroups 7A, 7B, 7C, and 7D. If the seed is not black and is more than 3 mm in diameter, it belongs in group 7C. Check the photographs or descriptions of the 4 species in group 7C to determine if the unknown seed matches one of them.

All 280 species are included in least one subgroup. Seeds with several conspicuous features are included in several groups. For example, *Bidens bipinnata* could be described as having barbs, bristles, or spines (1), as very large (17), as elongate (19), or as very elongate (20). Any of these groups could be used equally well as a starting point, since *Bidens bipinnata* is included in a subgroup of each.

The most difficult distinctions are between ovate and elliptical shapes and between ovoid and ellipsoid three-dimensional forms. Refer to the Illustrated Glossary for these and other shapes. It may be necessary to measure a seed; if so, a millimeter ruler and a 5× magnification hands lens are usually adequate.

Groups
1. Barbs, bristles, or spines present
2. Distinct lengthwise ribs or crosswise ridges present
3. Distinct surface texture but not ribs or ridges
4. Apical collar present
5. Grass floret, with 2 to several bracts enclosing a grain
6. Embryo evidently curved, folded, or coiled

7. Spherical form
8. Biconvex form, lenslike
9. Reniform
10. Trigonous form
11. Sectorform
12. Ovoid or obovoid form (not ellipsoid)
13. Angular form (other than trigonous)
14. Chiplike form
15. Form none of the above or highly irregular
16. Very small (no more than 0.6 mm long)
17. Very large (at least 8 mm long)
18. Very thin
19. Elongate (more than 2× long as wide; not more than 5× long as wide)
20. Very elongate (at least 6× long as wide)
21. Outline circular
22. Outline elliptical (not ovate or obovate)

Subgroups

1. Barbs, bristles, or spines
   1A. More than 5 mm long
      198. *Bidens cernua*, Nodding beggar-ticks
      200. *Bidens vulgata*, Tall beggar-ticks
      250. *Xanthium spinosum*, Spiny cocklebur
      251. *Xanthium strumarium*, Cocklebur

   1B. Less than 5 mm long; many curved spines
      190. *Ambrosia grayi*, Bur ragweed
      162. *Cynoglossum officinale*, Hound's tongue
      140. *Daucus carota*, Queen Anne's lace
      184. *Galium aparine*, Catchweed bedstraw

   1C. Less than 5 mm long; does not have many curved spines
      252. *Aegilops cylindrica*, Jointed goat grass
      197. *Bidens bipinnata*, Spanish needles
      199. *Bidens frondosa*, Devil's beggar-ticks
      234. *Lactuca saligna*, Willow-leaved lettuce
      235. *Lactuca serriola*, Prickly lettuce

2. Distinct ribs or ridges
   2A. Specifically, lengthwise ribs
      187. *Dipsacus fullonum*, Common teasel
      234. *Lactuca saligna*, Willow-leaved lettuce

235. *Lactuca serriola*, Prickly lettuce
242. *Solidago rigida*, Stiff goldenrod
248. *Vernonia baldwinii*, Western ironweed
249. *Vernonia fasciculata*, Ironweed

2B. Specifically, crosswise ridges
237. *Onopordum acanthium*, Scotch thistle
135. *Oxalis dillenii*, Gray-green wood sorrel
136. *Oxalis stricta*, Yellow wood sorrel
274. *Setaria faberi*, Giant foxtail
275. *Setaria glauca*, Yellow foxtail

2C. Specifically, concentric ridges
265. *Eleusine indica*, Goosegrass
28. *Mollugo verticillata*, Carpetweed
86. *Thlaspi arvense*, Pennycress

3. Distinct surface texture
3A. Specifically, papillate
29. *Cerastium vulgatum*, Mouse-ear chickweed
27. *Portulaca oleracea*, Purslane
31. *Saponaria officinalis*, Soapwort
32. *Silene antirrhina*, Sleepy catchfly

3B. Specifically, tuberculate
123. *Euphorbia cyathophora*, Fire-on-the-mountain
124. *Euphorbia dentata*, Toothed spurge
51. *Hibiscus trionum*, Flower-of-an-hour
30. *Holosteum umbellatum*, Jagged chickweed
146. *Hyoscyamus niger*, Henbane
163. *Lithospermum arvense*, Corn gromwell
8. *Mirabilis nyctaginea*, Wild four-o'clock
33. *Stellaria media*, Common chickweed
34. *Vaccaria pyramidata*, Cow-cockle

3C. Specifically, reticulate
50. *Abutilon theophrasti*, Velvet-leaf
2. *Argemone polyanthemos*, Prickly poppy
3. *Argemone squarrosa*, Hedgehog prickly poppy
145. *Datura stramonium*, Jimson weed
49. *Hypericum perforatum*, Common St. John's-wort
182. *Proboscidea louisianica*, Devil's claw
152. *Solanum rostratum*, Buffalo bur

3D. Specifically, knobby
    199. *Bidens frondosa*, Devil's beggar-ticks
    245. *Taraxacum laevigatum*, Red-seeded dandelion
    246. *Taraxacum officinale*, Common dandelion
    247. *Tragopogon dubius*, Goat's beard
    177. *Verbascum blattaria*, Moth mullein
    178. *Verbascum thapsus*, Common mullein

4. Apical collar
  4A. Shiny; persistent pappus
    196. *Aster ericoides*, Heath aster
    203. *Centaurea cyanus*, Cornflower
    241. *Solidago gigantea*, Late goldenrod
    242. *Solidago rigida*, Stiff goldenrod

  4B. Shiny; no pappus; no more than 4 mm long
    201. *Carduus acanthoides*, Plumeless thistle
    208. *Cirsium arvense*, Canada thistle
    209. *Cirsium flodmanii*, Flodman's thistle
    212. *Cirsium vulgare*, Bull thistle
    225. *Helianthus ciliaris*, Texas blueweed
    243. *Sonchus arvensis*, Field sow thistle

  4C. Shiny; no pappus; more than 4 mm long
    207. *Cirsium altissimum*, Tall thistle
    210. *Cirsium ochrocentrum*, Yellow-spine thistle
    211. *Cirsium undulatum*, Wavyleaf thistle

  4D. Persistent pappus of plumose bristles
    203. *Centaurea cyanus*, Cornflower
    205. *Centaurea repens*, Russian knapweed
    218. *Eupatorium altissimum*, Tall eupatorium
    232. *Kuhnia eupatorioides*, False boneset

  4E. Persistent pappus of scales
    206. *Cichorium intybus*, Chicory
    216. *Dyssodia papposa*, Fetid marigold
    221. *Gutierrezia dracunculoides*, Broomweed
    222. *Helenium amarum*, Bitter sneezeweed
    223. *Helenium autumnale*, Sneezeweed
    227. *Heliopsis helianthoides*, False sunflower

4F. Dull; persistent pappus of bristles; very thin cross section
    233. *Lactuca canadensis*, Wild lettuce
    235. *Lactuca serriola*, Prickly lettuce
    244. *Sonchus asper*, Prickly sow thistle

4G. Dull; persistent pappus of bristles; terete to compressed, but not very thin
    213. *Conyza canadensis*, Horseweed
    214. *Conyza ramosissima*, Spreading fleabane
    245. *Taraxacum laevigatum*, Red-seeded dandelion
    246. *Taraxacum officinale*, Common dandelion
    248. *Vernonia baldwinii*, Western ironweed
    249. *Vernonia fasciculata*, Ironweed

5. Grass floret
5A. Elongate (more than 3× long as wide); less than 5 mm long
    261. *Digitaria ciliaris*, Southern crabgrass
    263. *Digitaria sanguinalis*, Hairy crabgrass
    268. *Festuca octoflora*, Six-weeks fescue

5B. Elongate (more than 3× long as wide); more than 5 mm long; with 1 to several long awns
    252. *Aegilops cylindrica*, Jointed goat grass
    269. *Hordeum pusillum*, Little barley
    266. *Elymus canadensis*, Canada wild rye
    257. *Bromus secalinus*, Cheat
    258. *Bromus tectorum*, Downy brome

5C. Elongate (more than 3× long as wide); more than 5 mm long; with short awns or awnless
    252. *Aegilops cylindrica*, Jointed goat grass
    253. *Agropyron repens*, Quackgrass
    254. *Agropyron smithii*, Western wheat grass
    255. *Bromus inermis*, Smooth brome
    273. *Panicum virgatum*, Switchgrass

5D. Outline round, ovate, or elliptical; cross section plano-convex
    262. *Digitaria ischaemum*, Smooth crabgrass
    264. *Echinochloa crus-galli*, Barnyard grass

274. *Setaria faberi*, Giant foxtail
275. *Setaria glauca*, Yellow foxtail
276. *Setaria verticillata*, Bristly foxtail
277. *Setaria viridis*, Green foxtail

5E. Outline round, ovate, or elliptical; cross section elliptical or triangular
    260. *Cynodon dactylon*, Bermuda grass
    270. *Panicum capillare*, Common witchgrass
    271. *Panicum dichotomiflorum*, Fall panicum
    272. *Panicum miliaceum*, Broom-corn millet
    278. *Sorghum bicolor*, Shattercane
    279. *Sorghum halepense*, Johnson-grass

6. Embryo evidently curved, folded, or coiled
  6A. Elongate
    66. *Camelina microcarpa*, Small-seeded false flax
    67. *Camelina sativa*, Gold-of-pleasure
    68. *Capsella bursa-pastoris*, Shepherd's purse
    72. *Descurainia pinnata*, Tansy mustard
    73. *Descurainia sophia*, Flixweed
    74. *Erysimum cheiranthoides*, Wormseed wallflower
    75. *Erysimum repandum*, Bushy wallflower
    76. *Lepidium campestre*, Field peppergrass
    83. *Sisymbrium altissimum*, Tumbling mustard
    84. *Sisymbrium loeselii*, Tall hedge mustard
    85. *Sisymbrium officinale*, Hedge mustard

  6B. Not elongate; yellow, brown, or orange
    61. *Barbarea vulgaris*, Winter cress
    62. *Berteroa incana*, Hoary false alyssum
    69. *Cardaria draba*, Hoary cress
    70. *Chorispora tenella*, Blue mustard
    71. *Conringia orientalis*, Hare's-ear mustard
    77. *Lepidium densiflorum*, Peppergrass
    78. *Lepidium perfoliatum*, Clasping peppergrass
    79. *Lepidium virginicum*, Virginia peppergrass
    16. *Monolepis nuttalliana*, Poverty weed
    80. *Nasturtium officinale*, Watercress
    81. *Rorippa palustris*, Bog yellow cress
    86. *Thlaspi arvense*, Pennycress

6C. Embryo distinctly coiled; more or less round in form
  9. *Atriplex subspicata*, Spearscale
  28. *Mollugo verticillata*, Carpetweed
  27. *Portulaca oleracea*, Purslane
  17. *Salsola collina*, Tumbleweed
  18. *Salsola iberica*, Russian-thistle

6D. Not elongate; black or red-black
  60. *Cleome serrulata*, Rocky Mountain bee plant
  26. *Froelichia floridana*, Field snake-cotton
  15. *Kochia scoparia*, Fireweed

7. Spherical
  7A. Black; more than 3 mm in diameter
    132. *Cardiospermum halicacabum,* Common balloon vine
    109. *Vicia villosa*, Hairy vetch

  7B. Black; no more than 3 mm in diameter
    2. *Argemone polyanthemos*, Prickly poppy
    3. *Argemone squarrosa*, Hedgehog prickly poppy
    161. *Ellisia nyctelea*, Waterpod
    123. *Euphorbia cyathophora*, Fire-on-the-mountain
    34. *Vaccaria pyramidata*, Cow-cockle

  7C. Not black; more than 3 mm in diameter
    121. *Croton texensis*, Texas croton
    184. *Galium aparine*, Catchweed bedstraw
    130. *Tragia betonicifolia*, Nettleleaf noseburn
    131. *Tragia ramosa*, Noseburn

  7D. Not black; no more than 3 mm in diameter
    63. *Brassica juncea*, Indian mustard
    64. *Brassica kaber*, Charlock
    184. *Galium aparine*, Catchweed bedstraw

8. Biconvex form
    19. *Amaranthus albus*, Tumble pigweed
    20. *Amaranthus graecizans*, Prostrate pigweed
    21. *Amaranthus hybridus*, Slender pigweed
    22. *Amaranthus palmeri*, Palmer's pigweed

23. *Amaranthus retroflexus*, Rough pigweed
24. *Amaranthus rudis*, Water-hemp
25. *Amaranthus spinosus*, Spiny pigweed
10. *Chenopodium album*, Lamb's quarters
11. *Chenopodium berlandieri*, Pitseed goosefoot
12. *Chenopodium gigantospermum*, Maple-leaved goosefoot
13. *Chenopodium pratericola*, Dryland goosefoot
14. *Cycloloma atriplicifolium*, Tumble ringwing
4. *Corydalis aurea*, Golden corydalis

9. Reniform
9A. Smooth surface
91. *Astragalus canadensis*, Canada milk-vetch
92. *Astragalus mollissimus*, Woolly locoweed
99. *Medicago minima*, Small bur-clover
100. *Medicago sativa*, Alfalfa

9B. Tuberculate or papillate surface
50. *Abutilon theophrasti*, Velvet-leaf
31. *Saponaria officinalis*, Soapwort
32. *Silene antirrhina*, Sleepy catchfly

10. Trigonous
10A. Black
383. *Polygonum convolvulus*, Wild buckwheat
41. *Polygonum persicaria*, Lady's thumb
42. *Polygonum ramosissimum*, Bush knotweed

10B. Brown, orange-brown, or red-brown
36. *Polygonum arenastrum*, Knotweed
42. *Polygonum ramosissimum*, Bush knotweed
44. *Rumex acetosella*, Sheep sorrel
46. *Rumex crispus*, Curly dock
47. *Rumex mexicanus*, Willow-leaved dock
48. *Rumex patientia*, Patience dock

11. Sectorform
11A. More than 4 mm long
154. *Calystegia sepium*, Hedge bindweed
118. *Croton capitatus*, Woolly croton
156. *Ipomoea hederacea*, Ivyleaf morning-glory

157. *Ipomoea lacunosa*, White morning-glory
158. *Ipomoea purpurea*, Common morning-glory

11B. Less than 4 mm long; yellow
   29. *Cerastium vulgatum*, Mouse-ear chickweed
  159. *Cuscuta indecora*, Large alfalfa dodder
  160. *Cuscuta pentagona*, Field dodder
  171. *Salvia reflexa*, Lance-leaved sage
  134. *Tribulus terrestris*, Puncture vine
  165. *Verbena hastata*, Blue vervain
  167. *Verbena urticifolia*, Nettle-leaved vervain

11C. Less than 4 mm long; black
  215. *Coreopsis tinctoria*, Plains coreopsis
  145. *Datura stramonium*, Jimson weed
  52. *Malva neglecta*, Common mallow
  54. *Malvastrum hispidum*, Yellow false-mallow

11D. Less than 4 mm long; brown or red-brown
  155. *Convolvulus arvensis*, Field bindweed
  215. *Coreopsis tinctoria*, Plains coreopsis
  168. *Glecoma hederacea*, Ground ivy
  169. *Hedeoma hispidum*, Rough false pennyroyal
  170. *Lamium amplexicaule*, Henbit
  53. *Malva rotundifolia*, Round-leaf mallow
  55. *Sida spinosa*, Prickly sida
  164. *Verbena bracteata*, Prostrate vervain
  166. *Verbena stricta*, Hoary vervain

12. Ovoid or obovoid
12A. Black; smooth and shiny; less than 2 mm long
  19. *Amaranthus albus*, Tumble pigweed
  20. *Amaranthus graecizans*, Prostrate pigweed
  21. *Amaranthus hybridus*, Slender pigweed
  22. *Amaranthus palmeri*, Palmer's pigweed
  23. *Amaranthus retroflexus*, Rough pigweed

12B. Black; smooth and shiny; at least 2 mm long
  89. *Hoffmanseggia glauca*, Indian rush-pea
  37. *Polygonum bicorne*, Pink smartweed
  41. *Polygonum persicaria*, Lady's thumb

12C. Black or very dark; dull
    116. *Acalypha rhomboidea*, Rhombic copperleaf
    199. *Bidens frondosa*, Devil's beggar-ticks
    5. *Cannabis sativa*, Marijuana
    122. *Euphorbia corollata*, Flowering spurge
    231. *Iva xanthifolia*, Marsh elder
    59. *Sicyos angulata*, Bur cucumber

12D. Black or very dark; reticulate, tuberculate, or papillate surface
    60. *Cleome serrulata*, Rocky Mountain bee plant
    123. *Euphorbia cyathophora*, Fire-on-the mountain
    124. *Euphorbia dentata*, Toothed spurge
    51. *Hibiscus trionum*, Flower-of-an-hour
    27. *Portulaca oleracea*, Purslane
    182. *Proboscidea louisianica*, Devil's claw
    152. *Solanum rostratum*, Buffalo bur

12E. Mottled or streaked; less than 3 mm long
    116. *Acalypha rhomboidea*, Rhombic copperleaf
    124. *Euphorbia dentata*, Toothed spurge
    170. *Lamium amplexicaule*, Henbit
    97. *Lotus corniculatus*, Bird's-foot trefoil
    102. *Melilotus officinalis*, Yellow sweet clover
    106. *Trifolium campestre*, Low hop-clover
    107. *Trifolium pratense*, Red clover

12F. Mottled or streaked; more than 3 mm long
    191. *Ambrosia psilostachya*, Western ragweed
    118. *Croton capitatus*, Woolly croton
    237. *Onopordum acanthium*, Scotch thistle

12G. Green; hilum or attachment point on margin
    94. *Desmodium illinoense*, Illinois tickclover
    95. *Lespedeza cuneata*, Sericea lespedeza
    97. *Lotus corniculatus*, Bird's-foot trefoil
    98. *Medicago lupulina*, Black medick
    100. *Medicago sativa*, Alfalfa
    101. *Melilotus alba*, White sweet clover
    102. *Melilotus officinalis*, Yellow sweet clover

12H. Green; hilum or attachment point at base
    190. *Ambrosia grayi*, Bur ragweed
    5. *Cannabis sativa*, Marijuana
    125. *Euphorbia esula*, Leafy spurge
    127. *Euphorbia marginata*, Snow-on-the-mountain
    110. *Gaura longiflora*, Large-flowered gaura
    111. *Gaura parviflora*, Velvety gaura

12I. Yellow
    57. *Cucurbita foetidissima*, Buffalo-gourd
    101. *Melilotus alba*, White sweet clover
    102. *Melilotus officinalis*, Yellow sweet clover
    273. *Panicum virgatum*, Switchgrass
    148. *Physalis longifolia*, Common ground cherry
    149. *Solanum carolinense*, Carolina horse-nettle
    107. *Trifolium pratense*, Red clover
    108. *Trifolium repens*, White clover
    6. *Urtica dioica*, Stinging nettle

12J. Gray or white
    114. *Acalypha monococca*, Slender copperleaf
    115. *Acalypha ostryaefolia*, Hop-hornbeam copperleaf
    116. *Acalypha rhomboidea*, Rhombic copperleaf
    117. *Acalypha virginica*, Virginia copperleaf
    120. *Croton monanthogynus*, One-seeded croton
    122. *Euphorbia corollata*, Flowering spurge
    125. *Euphorbia esula*, Leafy spurge
    127. *Euphorbia marginata*, Snow-on-the-mountain

12K. Reddish or purplish
    114. *Acalypha monococca*, Slender copperleaf
    115. *Acalypha ostryaefolia*, Hop-hornbeam copperleaf
    19. *Amaranthus albus*, Tumble pigweed
    110. *Gaura longiflora*, Large-flowered gaura
    95. *Lespedeza cuneata*, Sericea lespedeza
    107. *Trifolium pratense*, Red clover

13. Angular form
    13A. No more than 2 mm long; with ridges, wrinkles, or papillae
        29. *Cerastium vulgatum*, Mouse-ear chickweed

126. *Euphorbia maculata*, Spotted spurge
128. *Euphorbia nutans*, Eyebane
177. *Verbascum blattaria*, Moth mullein
178. *Verbascum thapsus*, Common mullein

13B. No more than 2 mm long; surface finely textured
112. *Oenothera biennis*, Common evening primrose
113. *Oenothera laciniata*, Cut-leaved evening primrose
173. *Plantago major*, Common plantain
174. *Plantago patagonica*, Patagonian plantain
82. *Rorippa sinuata*, Spreading yellow cress
84. *Sisymbrium loeselii*, Tall hedge mustard

13C. More than 3 mm long
88. *Cassia chamaecrista*, Showy partridge pea
187. *Dipsacus fullonum*, Common teasel
26. *Froelichia floridana*, Field snake-cotton
104. *Strophostyles helvola*, Wild bean
105. *Strophostyles leiosperma*, Slick seed bean

14. Chiplike form
204. *Centaurea maculosa*, Spotted knapweed
273. *Panicum virgatum*, Switchgrass
175. *Plantago rugelii*, Rugel's plantain
150. *Solanum eleagnifolium*, Silver-leaf nightshade
151. *Solanum ptycanthum*, Black nightshade
179. *Veronica agrestis*, Field speedwell
180. *Veronica arvensis*, Corn speedwell
181. *Veronica peregrina*, Purslane speedwell

15. Form none of the above
   15A. No more than 2 mm long
15. *Kochia scoparia*, Fireweed
1. *Ranunculus abortivus*, Early wood buttercup
17. *Salsola collina*, Tumbleweed
18. *Salsola iberica*, Russian-thistle
56. *Viola rafinesquii*, Johnny-jump-up

   15B. 2–4 mm long
189. *Ambrosia artemisiifolia*, Common ragweed
138. *Cicuta maculata*, Water hemlock
139. *Conium maculatum*, Poison hemlock

220. *Grindelia squarrosa*, Curly top gumweed
225. *Helianthus ciliaris*, Texas blueweed
228. *Heterotheca latifolia*, Camphor weed
229. *Iva annua*, Sump weed
230. *Iva axillaris*, Poverty weed
231. *Iva xanthifolia*, Marsh elder
103. *Oxytropis lambertii*, Purple locoweed

15C. At least 4 mm long
190. *Ambrosia grayi*, Bur ragweed
191. *Ambrosia psilostachya*, Western ragweed
192. *Ambrosia trifida*, Giant ragweed
183. *Diodia teres*, Rough buttonweed
224. *Helianthus annuus*, Common sunflower
226. *Helianthus petiolaris*, Plains sunflower

16. Very small
267. *Eragrostis cilianensis*, Stinkgrass
28. *Mollugo verticillata*, Carpetweed
81. *Rorippa palustris*, Bog yellow cress

17. Very large
17A. Outline long and narrow
253. *Agropyron repens*, Quackgrass
254. *Agropyron smithii*, Western wheat grass
197. *Bidens bipinnata*, Spanish needles
255. *Bromus inermis*, Smooth brome
256. *Bromus japonicus*, Japanese brome
257. *Bromus secalinus*, Cheat
258. *Bromus tectorum*, Downy brome
247. *Tragopogon dubius*, Goat's beard

17B. Outline ovate or oblong, not very narrow
252. *Aegilops cylindrica*, Jointed goat grass
200. *Bidens vulgata*, Tall beggar-ticks
57. *Cucurbita foetidissima*, Buffalo-gourd
58. *Echinocystis lobata*, Wild cucumber
182. *Proboscidea louisianica*, Devil's claw
59. *Sicyos angulatus*, Bur cucumber
250. *Xanthium spinosum*, Spiny cocklebur
251. *Xanthium strumarium*, Cocklebur

18. Very thin
  18A. Smooth or finely textured; orange-brown
    141. *Asclepias subverticillata*, Poison milkweed
    142. *Asclepias syriaca*, Common milkweed
    143. *Asclepias verticillata*, Whorled milkweed
    144. *Cynanchum laeve*, Sand vine

  18B. Smooth or finely textured; gray or light colored
    196. *Aster ericoides*, Heath aster
    228. *Heterotheca latifolia*, Camphor weed
    238. *Silphium integrifolium*, Whole-leaf rosin-weed
    239. *Silphium laciniatum*, Compass plant
    240. *Silphium perfoliatum*, Cup plant

  18C. Ribbed lengthwise
    193. *Arctium minus*, Common burdock
    200. *Bidens vulgata*, Tall beggar-ticks
    255. *Bromus inermis*, Smooth brome
    233. *Lactuca canadensis*, Wild lettuce
    234. *Lactuca saligna*, Willow-leaved lettuce

  18D. With crosswise ridges
    135. *Oxalis dillenii*, Gray-green wood sorrel
    136. *Oxalis stricta*, Yellow wood sorrel

19. Elongate (more than 2× long as wide)
  19A. More than 5 mm long; a grass floret
    254. *Agropyron smithii*, Western wheat grass
    255. *Bromus inermis*, Smooth brome
    257. *Bromus secalinus*, Cheat
    269. *Hordeum pusillum*, Little barley

  19B. More than 5 mm long; apical collar present
    193. *Arctium minus*, Common burdock
    207. *Cirsium altissimum*, Tall thistle
    211. *Cirsium undulatum*, Wavyleaf thistle

  19C. More than 5 mm long; spines, barbs, or bristles present
    198. *Bidens cernua*, Nodding beggar-ticks
    200. *Bidens vulgata*, Tall beggar-ticks
    250. *Xanthium spinosum*, Spiny cocklebur
    251. *Xanthium strumarium*, Cocklebur

19D. Less than 5 mm long; a grass floret
    260. *Cynodon dactylon*, Bermuda grass
    261. *Digitaria ciliaris*, Southern crabgrass
    262. *Digitaria ischaemum*, Smooth crabgrass
    263. *Digitaria sanguinalis*, Hairy crabgrass
    264. *Echinochloa crus-galli*, Barnyard grass
    268. *Festuca octoflora*, Six-weeks fescue
    273. *Panicum virgatum*, Switchgrass
    276. *Setaria verticillata*, Bristly foxtail

19E. Less than 5 mm long; yellow or light colored; not a grass floret; no apical collar
    30. *Holosteum umbellatum*, Jagged chickweed
    146. *Hyoscyamus niger*, Henbane
    49. *Hypericum perforatum*, Common St. John's-wort

19F. Less than 5 mm long; with curved spines or bristles
    190. *Ambrosia grayi*, Bur ragweed
    162. *Cynoglossum officinale*, Hound's tongue
    140. *Daucus carota*, Queen Anne's lace
    184. *Galium aparine*, Catchweed bedstraw

19G. Less than 5 mm long; with 1 to many straight awns or barbs
    252. *Aegilops cylindrica*, Jointed goat grass
    197. *Bidens bipinnata*, Spanish needles
    199. *Bidens frondosa*, Devil's beggar-ticks
    234. *Lactuca saligna*, Willow-leaved lettuce
    235. *Lactuca serriola*, Prickly lettuce

19H. Less than 2 mm long; yellow; no awns or barbs
    259. *Chloris verticillata*, Windmill grass
    213. *Conyza canadensis*, Horseweed
    214. *Conyza ramosissima*, Spreading fleabane
    217. *Erigeron strigosus*, Daisy fleabane
    75. *Erysimum repandum*, Bushy wallflower

19I. 2–5 mm long; yellow; no awns or barbs
    202. *Carduus nutans*, Musk thistle
    207. *Cirsium altissimum*, Tall thistle
    208. *Cirsium arvense*, Canada thistle
    261. *Digitaria ciliaris*, Southern crabgrass

234. *Lactuca saligna*, Willow-leaved lettuce
273. *Panicum virgatum*, Switchgrass
276. *Setaria verticillata*, Bristly foxtail

19J. Less than 5 mm long; apex acute or acuminate; not a floret; no awns or barbs
    259. *Chloris verticillata*, Windmill grass
    129. *Euphorbia prostrata*, Creeping spurge
    49. *Hypericum perforatum*, Common St. John's-wort
    233. *Lactuca canadensis*, Wild lettuce
    234. *Lactuca saligna*, Willow-leaved lettuce
    235. *Lactuca serriola*, Prickly lettuce
    245. *Taraxacum laevigatum*, Red-seeded dandelion
    246. *Taraxacum officinale*, Common dandelion

19K. Less than 5 mm long; embryo evidently curved or folded
    66. *Camelina microcarpa*, Small-seeded false flax
    67. *Camelina sativa*, Gold-of-pleasure
    68. *Capsella bursa-pastoris*, Shepherd's purse
    72. *Descurainia pinnata*, Tansy mustard
    73. *Descurainia sophia*, Flixweed
    74. *Erysimum cheiranthoides*, Wormseed wallflower
    75. *Erysimum repandum*, Bushy wallflower
    76. *Lepidium campestre*, Field peppergrass
    85. *Sisymbrium officinale*, Hedge mustard

19L. Less than 5 mm long; smooth and shiny; not a grass floret
    90. *Amorpha canescens*, Lead plant
    201. *Carduus acanthoides*, Plumeless thistle
    202. *Carduus nutans*, Musk thistle
    207. *Cirsium altissimum*, Tall thistle
    208. *Cirsium arvense*, Canada thistle
    212. *Cirsium vulgare*, Bull thistle
    93. *Coronilla varia*, Crown vetch
    172. *Plantago lanceolata*, Buckhorn plantain

19M. Less than 5 mm long; rough surface
    129. *Euphorbia prostrata*, Creeping spurge
    49. *Hypericum perforatum*, Common St. John's-wort
    175. *Plantago rugelii*, Rugel's plantain
    242. *Solidago rigida*, Stiff goldenrod

245. *Taraxacum laevigatum*, Red-seeded dandelion
246. *Taraxacum officinale*, Common dandelion
248. *Vernonia baldwinii*, Western ironweed
249. *Vernonia fasciculata*, Ironweed

19N. Less than 5 mm long; persistent pappus of bristles
196. *Aster ericoides*, Heath aster
204. *Centaurea maculosa*, Spotted knapweed
213. *Conyza canadensis*, Horseweed
214. *Conyza ramosissima*, Spreading fleabane
242. *Solidago rigida*, Stiff goldenrod
248. *Vernonia baldwinii*, Western ironweed
249. *Vernonia fasciculata*, Ironweed

19O. Less than 5 mm long; apical collar present; pappus either deciduous or absent; dark colored
212. *Cirsium vulgare*, Bull thistle
221. *Gutierrezia dracunculoides*, Broomweed
227. *Heliopsis helianthoides*, False sunflower
244. *Sonchus asper*, Prickly sow thistle

19P. Less than 5 mm long; with apical collar; pappus either deciduous or absent; light colored
188. *Achillea millefolium*, Yarrow
195. *Artemisia ludoviciana*, White sage
201. *Carduus acanthoides*, Plumeless thistle
202. *Carduus nutans*, Musk thistle
207. *Cirsium altissimum*, Tall thistle
208. *Cirsium arvense*, Canada thistle
217. *Erigeron strigosus*, Daisy fleabane

20. Very elongate (at least 6× long as wide)
253. *Agropyron repens*, Quackgrass
197. *Bidens bipinnata*, Spanish needles
256. *Bromus japonicus*, Japanese brome
258. *Bromus tectorum*, Downy brome
266. *Elymus canadensis*, Canada wild rye
232. *Kuhnia eupatorioides*, False boneset
247. *Tragopogon dubius*, Goat's beard

21. Outline circular
21A. Margin with 1 or 2 notches
118. *Croton capitatus*, Woolly croton

52. *Malva neglecta*, Common mallow
53. *Malva rotundifolia*, Round-leaf mallow
28. *Mollugo verticillata*, Carpetweed
7. *Phytolacca americana*, Pokeweed

21B. Margin entire (lacking notches); no more than 2 mm long
9. *Atriplex subspicata*, Spearscale
160. *Cuscuta pentagona*, Field dodder
14. *Cycloloma atriplicifolium*, Tumble ringwing
16. *Monolepis nuttalliana*, Poverty weed
17. *Salsola collina*, Tumbleweed
33. *Stellaria media*, Common chickweed

21C. Margin entire (lacking notches); more than 2 mm long
9. *Atriplex subspicata*, Spearscale
145. *Datura stramonium*, Jimson weed
53. *Malva rotundifolia*, Round-leaf mallow
35. *Polygonum amphibium*, Swamp smartweed
40. *Polygonum pensylvanicum*, Pennsylvania smartweed

22. Outline elliptical
22A. At least 4 mm long
209. *Cirsium flodmanii*, Flodman's thistle
210. *Cirsium ochrocentrum*, Yellow-spine thistle
140. *Daucus carota*, Queen Anne's lace
58. *Echinocystis lobata*, Wild cucumber
278. *Sorghum bicolor*, Shattercane
279. *Sorghum halepense*, Johnson-grass

22B. Less than 4 mm long; smooth and shiny
90. *Amorpha canescens*, Lead plant
203. *Centaurea cyanus*, Cornflower
96. *Lespedeza stipulacea*, Korean lespedeza
102. *Melilotus officinalis*, Yellow sweet clover
270. *Panicum capillare*, Common witchgrass
271. *Panicum dichotomiflorum*, Fall panicum
272. *Panicum miliaceum*, Broom-corn millet
181. *Veronica peregrina*, Purslane speedwell
56. *Viola rafinesquii*, Johnny-jump-up

22C. Less than 4 mm long; dull
 62. *Berteroa incana*, Hoary false alyssum
 64. *Brasssica kaber*, Charlock
 65. *Brassica nigra*, Black mustard
 71. *Conringia orientalis*, Hare's-ear mustard
 228. *Heterotheca latifolia*, Camphor weed
 172. *Plantago lanceolata*, Buckhorn plantain
 174. *Plantago patagonica*, Patagonian plantain
 133. *Rhus glabra*, Smooth sumac
 81. *Rorippa palustris*, Bog yellow cress
 82. *Rorippa sinuata*, Spreading yellow cress
 171. *Salvia reflexa*, Lance-leaved sage
 153. *Solanum triflorum*, Cut-leaved nightshade
 244. *Sonchus asper*, Prickly sow thistle

22D. Less than 4 mm long; with ridges or other distinct texture
 138. *Cicuta maculata*, Water hemlock
 215. *Coreopsis tinctoria*, Plains coreopsis
 169. *Hedeoma hispidum*, Rough false pennyroyal
 228. *Heterotheca latifolia*, Camphor weed
 135. *Oxalis dillenii*, Gray-green wood sorrel
 136. *Oxalis stricta*, Yellow wood sorrel
 274. *Setaria faberi*, Giant foxtail
 275. *Setaria glauca*, Yellow foxtail
 276. *Setaria verticillata*, Bristly foxtail
 277. *Setaria viridis*, Green foxtail
 243. *Sonchus arvensis*, Field sow thistle

Seed Descriptions

The descriptions of the 280 species of seeds are arranged in systematic order. The sequence of families is that of Cronquist (1981); genera are presented alphabetically within the family. Each description includes the following information, in this order:

- Species number (according to the systematic order).
- Scientific name and authority.
- Common name.
- The unit of dispersal being described (a true seed or a fruit; if a fruit, what type).
- Shape and structure. The first paragraph describes outline shape, cross-section shape, and three-dimensional form. Any external structural features such as the hilum are described. Consult the Illustrated Glossary to determine the meaning of the descriptive terms for shape and form and the Glossary for terms for structures.
- Surface characteristics. The second paragraph describes the surface texture, sheen, color, and any other characteristics of the surface. Descriptive terms for surface texture are included in the Illustrated Glossary. Since the appearance of seed surfaces may vary greatly with the degree of magnification used, this information is included with some of the descriptions. When the magnification is not specified, it is between 4× and 10×.
- Size. The third paragraph gives the range of length, width, and thickness measurements obtained. Length is the longest dimension of the seed; width is a line perpendicular to the length, through the widest part; thickness is a line perpendicular to both the length and the width. Measurement of a number of individual seeds may yield a range of typical sizes,

but some individuals may be much larger or smaller than typical. In such a case, the measurement of the exception to the range is indicated in parentheses.

- Notes on the distribution of the plant; the types of weedy habitats where the plant is most often found; its soil, light, or moisture preferences; and its geographic range within the Great Plains. Weeds by their nature can be found almost anywhere and often are adapted to a broad range of habitats and conditions. Therefore these notes should be understood to indicate the most likely places to find the plants rather than to exclude the possibility of finding them elsewhere.

**Ranunculaceae
Buttercup Family**

1. *Ranunculus abortivus* L.

**Early wood buttercup,
small-flowered crowfoot**

Achene: Outline rounded triangular or ovate with 1 nearly straight edge. Cross section biconvex. Form ovoid, compressed, with thin, narrow margins. From the attachment point (at the small end of the achene) 2 thickened marginal ridges run most of the distance to the wide end, forming the straight edge. These ridges end in a minute hook-shaped style remnant that curves toward the wide end. The remainder of the margin is narrowly winged.

Surface smooth; glossy. Small shallow pits are visible with magnification. Light orange-brown, uneven in intensity.

Length 1.4–1.8 mm; width 1.2–1.4 mm; thickness 0.7 mm.

Gardens, pastures; moist places; E 1/3 GP.

**Papaveraceae
Poppy Family**

**2. *Argemone polyanthemos*
(Fedde) G. Ownbey**

Prickly poppy

Seed: Outline round, with 1 or 2 small pointed projections. Cross section round. Form nearly globose but with a thin linear seam projecting above the surface for about 1/3 the circumference. One end of the seam forms a blunt tooth, and the other end is a small extended tip ±0.2 mm long.

Surface covered with a reticulate pattern of sharp-edged ridges with smooth pits between them. This pattern is visible without magnification and is very distinct at 7×. The rows of pits appear to radiate from the small extended tip. The mesh looks like a membrane shrunk to the surface of the seed. Dull, but under magnification the pits are shiny. Dark gray to nearly black; the ridges and seam may be paler.

Diameter 1.8–2.2 mm.

Roadsides, pastures; sandy soils; S 2/3 GP.

3. *Argemone squarrosa* Greene   Hedgehog prickly poppy

Seed: Outline round with a distinct small tip and 2 blunt projections. Cross section round. Form nearly globose, with a toothlike projection, ±0.3–0.4 mm long. The hilum is next to the tooth. The hilum has an ovate rim that continues as a prominent ridge toward a point not quite opposite the tooth. Each end of this ridge forms a small blunt projection.

Surface covered with a conspicuous reticulate pattern of sharply defined ridges and smooth interspaces, visible without magnification; dull. Gray-brown, but when viewed at 15× magnification, the ridges are pale with fine dark central lines, and the interspaces are dark with a dense network of pale veins.

Diameter 2.3–2.6 mm.

Pastures, rangeland, roadsides; sandy soil; SW 1/4 GP.

Fumariaceae Fumitory Family

4. *Corydalis aurea* Willd.   Golden corydalis

Seed: Outline elliptical to nearly round, with 2 notches, one deep and one shallow, at one end. Cross section biconvex. Form elliptical, compressed, with a conspicuous notch near the basal end. The hilum is in a broad shallow notch at the center of the basal end, next to the larger notch. The hilum is ±0.3 mm in diameter, pale and dull.

Surface smooth; very shiny. At 10× or greater magnification there is a low-relief tuberculate or scalariform pattern that is more pronounced near the margin of each face. Black.

Length 1.5–1.7 mm; width 1.4–1.6 mm; thickness 0.9–1.0 mm.

Roadsides; moist areas, often gravelly or sandy soil; N 1/2 and S 1/4 GP.

Cannabaceae Hemp Family

5. *Cannabis sativa* L.   Marijuana, hemp

Achene: Outline short ovate with a truncate, sometimes extended base. Cross section biconvex but asymmetrical. Form ovoid, slightly compressed, with a blunt-tipped apex and a small base that extends downward. The dorsal and ventral faces meet in a narrow rim. The attachment area, at the base, is large, round, and surrounded by a thickened collar.

Surface smooth; slight sheen. Under magnification it has a parchmentlike luster. Patchy in color, ranging from dull green or greenish black to light brown or dark brown. With magnification fine dark streaks are visible in the paler areas, and there is a fine, light-colored network over the whole seed. This network appears to lie in a translucent surface layer.

Length (3.1) 3.6–4.2 mm; width (2.4) 2.7–3.3 mm; thickness 2.2–2.6 mm.

Pastures, roadsides; open areas; E 2/3 of Kansas, E 1/2 of Nebraska.

**Urticaceae
Nettle Family**

6. *Urtica dioica* L.   **Stinging nettle**

Achene: Outline ovate with a slightly extended base. Cross section narrowly biconvex. Form ovoid, compressed, with an acute tip at the small end and a narrow marginal wing all around. There is a small stalklike extension from the center of the wide end of the achene; this is the attachment point.

Surface finely textured; dull. Pale dull orange; the wing is paler.

Length 1.0–1.1 mm; width 0.7–0.8 mm; thickness 0.3–0.4 mm.

Pastures, fence rows; moist areas, often in shade; N 2/3 GP.

**Phytolaccaceae
Pokeweed Family**

7. *Phytolacca americana* L.   **Pokeweed, pokeberry**

Seed: Outline nearly circular, with a small extension and a V-shaped notch. Cross section elliptical. Form discoid, with rounded margins. There may be a depression in the center of each face. There is 1 thin, somewhat angular extension of the margin; next to the extension is the hilum, a broad notch partly filled with corky tissue.

Surface very smooth, very glossy. Black, with purplish undertones.

Length 2.8–2.9 mm; width 2.5–2.8 mm; thickness 1.3–1.4 mm.

Gardens, pastures; rich soil in partial shade; SE 1/4 GP.

**Nyctaginaceae
Four-O'Clock
Family**

8. *Mirabilis nyctaginea* (Michx.) MacM.   **Wild four-o'clock**

Anthocarp (dry accessory fruit): Outline obovate with a small truncate base and a short rectangular extension at the apex. Cross section approximately round; there are 5 angles, but they are obscured by hairs. Form clavate, with a truncate base. There are 5 long faces that meet in 5 lengthwise angles. There is a small cylindrical extension at the apex. The attachment area, at the small end, has a circular rim and a small point at the center.

Surface very rough, hairy; dull. There are large irregular tubercles on the angles. The faces have small tuberculate wrinkles, with fine striations between the wrinkles. There are abundant white multicellular hairs. Gray-brown overall; the ridges and wrinkles are yellow-brown, and the interspaces are dark brown.

Length 3.5–3.7 mm; diameter 1.7–1.9 mm.
Field margins, roadsides, pastures; good soils; throughout GP.

**Chenopodiaceae
Goosefoot Family**

9. *Atriplex subspicata* (Nutt.) Rydb.   **Spearscale, saltbush**

Seed: The seeds of this species are dimorphic; the 2 forms may be found on the same plant.

Smaller form: Outline nearly round, with 1 rounded extension that forms a notch and a tiny tip (style base) located about 1/3 the distance around the margin from the extension. Cross section elliptical. Form a thick disk with rounded margins. On each face there is a groove that runs from the notch toward the center. The membranous pericarp is often present.
    Surface smooth; shiny. Black. The pericarp is greenish, with a slight sheen and lengthwise striations, but allows the dark color of the seed to show.
    Diameter 1.4–1.6 mm; thickness 0.7–0.8 mm.

Larger form: Outline nearly round, with a distinct tapered extension. Cross section a narrow wedge. Form discoid, apparently a coil with one free end; the margin of the disk is thicker toward the inner end of the coil. The membranous pericarp is often present.
    Surface smooth; dull. With magnification it appears finely textured. Orange-brown to dark yellow-brown. The pericarp is thin, greenish; it allows the color of the seed to show.
    Length 2.6–3.0 mm; width 2.2–2.9 mm; thickness 0.9–1.2 mm.
    Waste places; saline or alkaline soil; throughout GP.

10. *Chenopodium album* L.   **Lamb's quarters**

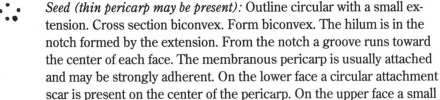

Seed (thin pericarp may be present): Outline circular with a small extension. Cross section biconvex. Form biconvex. The hilum is in the notch formed by the extension. From the notch a groove runs toward the center of each face. The membranous pericarp is usually attached and may be strongly adherent. On the lower face a circular attachment scar is present on the center of the pericarp. On the upper face a small style remnant projects slightly from the center of the pericarp.
    Surface (without the pericarp) nearly smooth and very shiny. Radial striations are visible at high magnification. Black. With the pericarp present, the surface is slightly roughened and radially striate. The pericarp is transparent, yellowish; it has a silky sheen.

Length 1.1–1.3 mm; width 1.0–1.1 mm; thickness 0.5–0.6 mm.
*Cultivated fields, gardens, roadsides, waste areas; disturbed soil;
E 1/2 GP.*

11. *Chenopodium berlandieri* Moq.   Pitseed goosefoot

Seed (with or without thin pericarp): Outline nearly circular. Cross section a rounded shape with one slightly convex edge; the other edge is strongly convex or nearly pointed. Form a short broad cone; the pointed end is at the base in fruit, and the top surface is slightly convex. Some seeds are more nearly biconvex. On the margin there is a small extension that forms a notch. On the upper face there is a slight groove from the notch to the center. On the lower face there is a wider groove. The seed is often covered by the thin, transparent, yellowish pericarp.

Surface (without the pericarp) smooth and shiny. A fine reticulate pattern is visible at 10× magnification. Black. When the pericarp is present the surface has a membranous sheen; a fine sharp reticulate pattern, arranged in radial lines, is visible with magnification. There is an opaque pale spot, the style base, in the center of the top face of the pericarp. On the lower face there is a dark spot in the center, at the attachment point. Yellowish gray, with the dark seed showing through the pericarp.

Diameter 1.3–1.4 mm; thickness 0.7–0.8 mm.
Cultivated fields, pastures, rangeland; disturbed soil; throughout GP.

12. *Chenopodium gigantospermum* Aellen   Maple-leaved goosefoot

Seed (with or without thin pericarp): Outline nearly circular, with a small curved extension that forms a small notch. In cross section there are 2 unequally convex sides, with narrow wings between them. Form biconvex with a thin border area around the circumference on each face; this border area is ±0.2 mm wide. On the edge are a notch and a rounded extension. The hilum is located in the notch. On the more convex face (the lower side in fruit) is a groove that runs from the notch to the center. All or part of the persistent pericarp may be present. On the lower face the pericarp has a round attachment scar. In the center of its upper face there is a small style base.

Surface (without the pericarp) smooth; shiny. At high magnification, the upper face appears slightly roughened, and the lower face has fine radial striations. Black. The pericarp is thin, yellowish, and nearly transparent, with radial striations on the lower face; with the pericarp present the apparent color of the seed is dull brown.

Length 2.2–2.4 mm; width 2.1–2.3 mm; thickness 1.1–1.2 mm.

The perianth may be persistent on the fruit. Five widely spaced sepals are folded over the rim of the fruit but do not completely enclose it.

Gardens, waste places; disturbed soil, often in shaded areas; throughout GP except SW.

13. *Chenopodium pratericola* Rydb.   Dryland goosefoot

Seed (with or without thin pericarp): Outline nearly circular with a small curved extension and a very small notch. In cross section there is 1 convex face and 1 face that is more strongly convex or nearly pointed. Form biconvex but unequally so; the lower face may be nearly conic. Hilum very small, circular, in the notch. On each face there is a groove that extends inward from the notch. The pericarp is readily separated from the seed but may be present. A tiny style remnant may be present on the upper face of the pericarp. The seed is exposed through a break in the pericarp at the center of the lower face.

Surface (without the pericarp) smooth; shiny. At 10× magnification fine radial wrinkles are visible; there are also faint parallel ridges on the small extension. Black. The pericarp is a thin translucent membrane, radially striate, with a slight sheen. With the pericarp present, the apparent color is dark yellowish gray.

Length 1.1–1.3 mm; width 1.0–2.2 mm; thickness 0.7–0.8 mm.
Rangeland, waste areas; dry soil; throughout GP.

14. *Cycloloma atriplicifolium* (Spreng.) Coult.   Tumble ringwing

Seed (with or without thin pericarp): Outline nearly circular; there may be a small notch. Cross section biconvex, but one face is much more strongly convex than the other. Form a disk with 1 distinctly convex face. On the convex face a shallow groove runs from the marginal notch to the center.

Surface smooth; shiny. Black. A membranous adherent pericarp is usually present. The pericarp is dull and papery and has fine radial striations. With the pericarp present, the apparent color is gray-black.

Diameter 1.5–1.8 mm, thickness 0.7–0.9 mm.

The seed is likely to be found still enclosed in the complete perianth, which is 2.2–2.5 mm in diameter. The lower side of the perianth is dark and cup-shaped with a ruffled marginal wing. The upper side has a dark brown center and a light brown ruffled wing. The 5 calyx parts are keeled and closely appressed to the upper face of the fruit but do not completely enclose it; a 5-armed star shape appears in the center.

Row crops; sandy soil; S 2/3 GP.

15. *Kochia scoparia* (L.) Schrad.   Fireweed, kochia

Seed: Outline pear shaped. Cross section a narrow triangle, with a notch in the short side. Form compressed ovoid, apparently folded. Two long tips meet to form the small end of the seed. A groove extends from the small end toward the center of each face. There is usually an ovate depressed area in the center of each face. One long margin of the seed is thicker than the other; the thicker margin has a groove along its outer face.

Surface very finely roughened; slight sheen. Dark red-brown to black; the thin edge of the seed is sometimes paler.

Length 1.1–2.1 mm; width 0.9–1.5 mm; thickness 0.5–0.8 mm.

Wheat, corn, sorghum, and alfalfa fields, waste areas, roadsides, pastures; disturbed soil; throughout GP.

16. *Monolepis nuttalliana* (R. & S.) Greene   Poverty weed

Seed (with or without thin pericarp): Outline nearly circular. Cross section a narrow ellipse. Form a disk with a thin rim. The seed is vertical in the fruit, and there is a notch in the margin at the basal attachment point. The U-shaped embryo can be distinguished on both faces, and the center of each face is slightly depressed. Part or all of the pericarp may be present.

Surface very finely textured; dull. Dark orange-brown. The pericarp is membranous, finely reticulate, and pale greenish in color. The rim of the pericarp is compressed. There may be a bifid style on the rim of the pericarp at the apex.

Diameter 1.1–1.3 mm; thickness 0.3–0.4 mm.

Pastures, waste areas; saline or alkaline soil; throughout GP.

17. *Salsola collina* Pall.   Tumbleweed

Seed: Outline roughly circular. Cross section round. Form somewhat conical, with the small end of the cone at the base in the fruit; the broad end of the cone is more or less concave. This cone is formed by the embryo, which consists of 2 spiral coils, one inside the other. A clinging membranous seed coat covers the coils. There is a style remnant at the center of the upper face.

Surface (of the seed coat) papery, striate; dull. Brown. The embryo is nearly smooth, semiglossy, olive green, but darker toward the center of the coils.

Diameter (1.2) 1.4–1.6 mm; thickness 0.7–1.1 mm. (Thickness measured from base to apex.)

Cultivated fields, roadsides; open areas; North Dakota and W central GP.

18. *Salsola iberica* Senn. & Pau.   Russian-thistle, tumbleweed

Seed: Outline obovate, with a pointed small end. Cross section round. Form conical with an oblique, pointed base. The embryo itself is coiled, and only a very thin membranous seed coat covers the embryo.

Surface (of the seed coat) papery; dull. Brown. The embryo is smooth, shiny, dark green in the center, fading to brownish green toward the outer coil.

Length 1.7–1.8 mm; thickness 1.2–1.4 mm.

Seed with perianth: The seed is often found still enclosed in all or part of the adherent papery perianth. The perianth has a flaring conical form, with an oblique flat apex. There is a style remnant in the center of the apex. The flat area has an irregular horizontal wing with 5 rounded teeth.

Roadsides, cultivated fields, especially wheat; dry soil; throughout GP.

Amaranthaceae Pigweed Family

19. *Amaranthus albus* L.   Tumble pigweed, tumbleweed

Seed: Outline ovate to nearly round. Cross section biconvex. Form ovoid, compressed, with 2 faces that meet in a narrow marginal rim. On each face there is a border area, ±0.1 mm wide, that is more compressed than the rest of the seed. Near the small end there is a small marginal notch; the hilum is in the notch.

Surface smooth; very shiny. At 10× magnification the border areas appear somewhat dull; at high magnification they are faintly reticulate. Black; less mature seeds reddish-black.

Length 1.0–1.1 mm; width 0.9–1.0 mm; thickness 0.6–0.7 mm.

Cultivated fields, rangeland, waste areas; dry soil; throughout GP.

20. *Amaranthus graecizans* L.   **Prostrate pigweed**

Seed: Outline ovate or sometimes nearly round. Cross section biconvex with sharply defined edges. Form ovoid, slightly compressed, with a distinct thin border on each face; the border is ±0.2 mm wide and 0.1 mm thick. The small end of the seed is truncate or slightly notched, with the hilum in the notch.

Surface very smooth; glassy luster. At high magnification, a very fine reticulate pattern is visible in the border area, with a similar faint pattern in the center of each face. Black.

Length 1.5–1.7 mm; width 1.3–1.5 mm; thickness ±0.7–0.8 mm.

Cropland, waste areas, roadsides; dry soil; throughout GP.

21. *Amaranthus hybridus* L.   **Slender pigweed**

Seed: Outline ovate. Cross section biconvex. Form ovoid, compressed, with 2 faces that meet in a rounded narrow rim. On each face there is a border that is slightly more compressed than the rest of the seed. Near the small end of the seed is a small, distinct, V-shaped notch that includes the hilum.

Surface smooth; shiny. The border areas are slightly dull. At 25× magnification a faint reticulate pattern, in concentric lines, is visible on the borders. Black, with red-brown undertones; sometimes the border areas are reddish.

Length 1.0–1.1 mm; width 0.8–0.9 mm; thickness 0.6–0.7 mm.

Corn, sorghum, soybean, and alfalfa fields, waste places; moist soil; E central GP.

22. *Amaranthus palmeri* S. Wats.   **Palmer's pigweed**

Seed: Outline ovate. Cross section biconvex. Form ovoid, compressed, with a small notch near the small end; the tiny hilum is in the notch. There is a slightly compressed border around each face.

Surface smooth, highly glossy; at magnifications of 7× or more, the border area appears duller than the center. At 10× magnification, the central area of each face appears very smooth and the border area slightly roughened. At 40×, the border appears to have a finely alveolate texture. Uniformly black.

Length 0.9–1.2 mm; width 0.8–0.9 mm; thickness 0.4–0.6 mm.

Corn, sorghum, soybean, and alfalfa fields, waste places; sandy soil; S 1/2 GP.

23. *Amaranthus retroflexus* L.   Rough pigweed

Seed: Outline ovate with a tiny tooth or notch just below the small end. Cross section biconvex. Form ovoid, compressed, with a narrow rim. The outer margins of the 2 faces appear slightly compressed, forming a border area ±0.1 mm wide. The hilum is small and round, located in the V-shaped notch; the lower edge of the notch sometimes projects as a small tooth.

Surface very smooth; very shiny with a glassy luster. Under high magnification the border area appears slightly roughened. Black.

Length 1.1–1.3 mm; width 0.9–1.1 mm; thickness 0.6–0.7 mm.

Corn, sorghum, soybean, and alfalfa fields, gardens, abandoned fields; good soils; throughout GP.

24. *Amaranthus rudis* Sauer   Water-hemp

Seed: Outline nearly circular. Cross section biconvex. Form a biconvex lens, with a slightly compressed border area on each face. The hilum is very small and is located in a marginal notch formed by 2 slightly extended tips; the rim of the hilum may appear as a tiny tooth.

Surface very smooth; glassy luster but the border areas are slightly dull. At 25× magnification, the borders are marked with a regular, faint scalariform pattern. Black, with undertones of red in the border areas.

Length 0.9 mm; width 0.8–0.9 mm, thickness 0.5–0.6 mm.

Waste areas, corn, sorghum, soybean, and alfalfa fields; moist soil; throughout GP except NW.

25. *Amaranthus spinosus* L.   Spiny pigweed

Seed: Outline ovate. Cross section biconvex. Form ovoid, compressed, with 2 faces that meet in a rounded rim. On each face there is a slightly compressed border. There is a tiny notch in the rim near the small end of the seed. The hilum is in the notch, and its edge projects as a small tooth.

Surface smooth, shiny. At 25× magnification a fine reticulate pattern, in concentric bands, is visible on the border area. Dark red-brown.

Length 0.9–1.1 mm; width 0.8–0.9 mm; thickness 0.5–0.6 mm.

Waste areas, feedlots, pastures; dry or irrigated soil; SE 1/4 GP.

26. *Froelichia floridana* (Nutt.) Moq.   Field snake-cotton

Seed: Outline rounded triangular. Cross section elliptical. Form a thick triangle, thinner toward one corner. The 2 faces of the triangle are slightly convex; they are joined by a broad band around the perimeter. The embryo is evidently folded, with 2 unequal tips that join to form the thinner corner of the seed.

Surface nearly smooth, with fine rounded wrinkles; shiny. Very dark red.

Length 1.5–1.7 mm; width 1.4–1.8 mm; thickness 1.0–1.4 mm.
Roadsides, waste areas; dry sandy soil; S 1/2 GP.

Portulacaceae
Purslane Family

27. *Portulaca oleracea* L.   **Purslane**

Seed: Outline ovate to rounded triangular, with a small notch on one side just below the small end. Cross section long ovate. Form ovoid, compressed, like a coil with the free end forming the small end of the ovoid form. Each face has a shallow groove, extending from the small end toward the center. The hilum area, located in the notch, is covered by a conspicuous piece of thin, light-colored tissue, placed crosswise to the edge of the seed.

Surface covered with distinct small tubercles; shiny. The tubercles are arranged in rows around the circumference of the seed. Dark red-brown to black.

Length 0.7–0.8 mm, width 0.5–0.7 mm; thickness 0.3–0.5 mm.
Fallow fields, cultivated fields, lawns, gardens; seasonally moist soil; throughout GP.

Molluginaceae
Carpetweed
Family

28. *Mollugo verticillata* L.   **Carpetweed**

Seed: Outline like a comma, with a small projection from the notch of the comma. Cross section ovate. Form ovoid, compressed, apparently a coil; thicker toward the outer edge of the coil. There are 5 to 7 distinct continuous narrow ridges around the circumference of the seed, from the narrow end to the notch, and several similar but fainter ridges concentric on each broad face. The hilum is a small conical tooth that projects ±0.1 mm from the center of the notch.

Surface conspicuously ridged; shiny. Orange-brown; the ridges and the small tooth are darker. The ridges appear translucent when seen with magnification.

Length 0.5–0.6 mm; width 0.4–0.6 mm; thickness ±0.4 mm.
Row crops; disturbed soil; S 1/2 GP but infrequent to the west.

Caryophyllaceae
Pink Family

29. *Cerastium vulgatum* L.   **Mouse-ear chickweed**

Seed: Outline roughly triangular. Cross section ovate. Form sectorlike, irregular, with some rounded and some angular edges. The round hilum, on one corner, is in a small notch formed by an incurved tip of tissue.

Surface papillate; many of the papillae are elongate, and they appear to radiate from the hilum area. When seen without magnification the surface appears dull, but at low magnification the papillae are shiny, and at high magnification they are transparent. Light orange-brown.

Length 0.6–0.7 mm; width 0.6–0.7 mm; thickness 0.3–0.5 mm.
Lawns; moist or shaded areas; E 1/3 GP.

30. *Holosteum umbellatum* L.   Jagged chickweed

Seed: Outline ovate. Cross section rounded triangular. Form ovoid, very compressed, bent; the dorsal face has a lengthwise groove and an extended, folded tip that continues as a ridge, 0.1–0.2 mm wide, for most of the length of the ventral face. The inconspicuous hilum is near the end of the ridge.

Surface roughened with closely spaced protuberances that are more prominent on the dorsal face; dull. These protuberances appear round when viewed at 10× magnification, but at 30× they appear as many-rayed stars; on the ridge they are elongate. Pale orange-brown.

Length 0.8 mm; width 0.6–0.7 mm; thickness 0.4–0.5 mm. The ridge is ±0.2 mm thick.
Lawns; sandy or rocky soil; SE 1/4 GP.

31. *Saponaria officinalis* L.   Soapwort, bouncing Bet

Seed: Outline ovate, with a distinct small notch at the center of one side. Cross section a narrow wedge. Reniform, compressed, thinner toward the notched edge and often appearing somewhat wrinkled or irregular. The small indistinct hilum is in the notch.

Surface papillate; dull. The papillae are round to elongate and appear to radiate from the notch area. Under magnification they are shiny. Black.

Length 1.8–2.2 mm; width 1.4–1.8 mm; thickness 0.5–0.7 mm.
Roadsides, waste areas; moist soil; E 1/3 GP and North Dakota.

32. *Silene antirrhina* L.   Sleepy catchfly

Seed: Outline kidney shaped to semicircular. Cross section rounded oblong. Reniform, with 2 nearly flat broad faces and a slight notch. The hilum is an elliptical cavity lying crosswise in the notch.

Surface papillate; the papillae are small, sharply defined, and closely spaced. The papillae are aligned in regular rows around the entire outer edge, from one side of the hilum to the other; on the broad faces of the seed the papillae are uniformly spaced. The surface appears dull, but at 10× magnification the tips of the papillae are shiny. Black.

Length 0.6–0.7 mm; width 0.5–0.6 mm; thickness 0.4–0.5 mm.

Fields of small grains, waste places; sandy soil; E 1/2 GP.

33. *Stellaria media* (L.) Cyr.   Common chickweed

Seed: Outline nearly circular. Cross section wedge shaped. Form discoid with 2 flat faces that join the margin at nearly right angles; the disk is thicker toward one margin. The seed appears folded, with 2 small tips that meet at the margin on the thinner margin. A groove between the tips extends about 1/4 the distance to the center on each flat face.

Surface tuberculate; dull. The tubercles are arranged in rows around the marginal face and in concentric circles on each broad face. Straw colored to pale reddish brown. (The color darkens with maturity, but the seeds are often shed while still immature.)

Length 0.8–0.9 mm; width 0.8–0.9 mm; thickness 0.4 mm.

Fall-seeded small grains, lawns, gardens; moist or shaded areas; E 1/2 GP.

34. *Vaccaria pyramidata* Medic.   Cow-cockle

Seed: Outline round to rounded triangular. Cross section approximately round. Form globose, but with one extended area. The hilum is on the extension, and the seed appears variously compressed or shrunken near the hilum.

Surface tuberculate; dull. The tubercles are small and uniform, visible at 7× magnification. The tubercles are arranged in distinct rows on a broad band around the circumference, from one side of the hilum to the other; they are evenly spaced over the rest of the surface. Black with red-brown undertones.

Length 1.7–2.0 mm; width 1.5–1.7 mm; thickness 1.5–1.7 mm.

Spring wheat, roadsides, waste places; good soils; N 2/3 GP.

Polygonaceae
Buckwheat Family

35. *Polygonum amphibium* L. var. *emersum* Michx.   Swamp smartweed, water smartweed

Achene: Outline nearly round with a small apical extension. Cross section a rounded oblong. Form diskoid, with rounded margins and a small tip. Both faces may be flat, but often one face is slightly concave or appears to be creased lengthwise on its center line. The small round attachment point is at the center of the base, opposite the tip. Perianth remnants and a style remnant are often present.

Surface smooth; very shiny. Under magnification it appears to have very fine, slight wrinkles. Black, often with red-brown undertones.

Length 2.8–3.1 mm; width 2.7–3.0 mm; thickness 0.9–1.1 mm.

Cropland, roadside ditches; damp soil or shallow water; throughout GP.

36. *Polygonum arenastrum* Jord. ex Bor.   Knotweed

Achene: Outline ovate to teardrop shaped. Cross section triangular, usually with sides of 3 different lengths; the sides may be straight, convex, or concave. Form 3-sided and 3-angled, with the angles curved gradually to an acute apex and abruptly to a rounded base. Calyx remnants are usually present at the base. At the apex there may be 1 to 3 small teeth, remnants of the style base.

Surface smooth; somewhat shiny, angles glossy. At 7× magnification it appears minutely granular; at high magnification it appears shiny but finely papillate in a lengthwise pattern. Dark red-brown.

Length 2.0–2.4 mm; width 1.3–1.6 mm; thickness 0.8–1.0 mm.

Waste areas, lawns, roadsides; dry, compacted soil; throughout GP.

37. *Polygonum bicorne* Raf.   Pink smartweed

Achene: Outline ovate or broadly elliptical, with a blunt base and a small acute tip. Cross section a broad, rounded triangle. Form ovoid, compressed, with one flat side; the other side has a lengthwise or central convexity. The attachment point is at the center of the base, opposite the tip. It is round with a distinctly projecting rim. Style and perianth remnants may be present.

Surface smooth, very glossy. Dark red-brown to black; the attachment point is white.

Length 2.8–2.9 mm; width 2.2–2.3 mm; thickness 0.9–1.1 mm.

Roadside ditches; moist disturbed soil; S 2/3 GP.

38. *Polygonum convolvulus* L.   Wild buckwheat

Achene: Outline elliptical. Cross section triangular, with concave sides of 3 different lengths. Form ellipsoid, but 3-sided and 3-angled, broadest at about the middle, with a blunt base and narrow apex. The attachment point is at the center of the base. Perianth remnants and a style remnant may be present.

Surface smooth; dull but the angles are glossy. Under magnification the surface is finely reticulate or striate. Black; the attachment point is pale.

Length 3.5–4.0 mm; width (of one face) 2.4–2.7 mm.

Roadsides, fence rows, waste places, cultivated fields, especially winter wheat; open areas; throughout GP.

39. *Polygonum lapathifolium* L.   Pale smartweed

Achene: Outline ovate to teardrop shaped, with a distinct acute tip. Cross section rounded oblong. Form ovoid, strongly compressed, with thick, rounded edges. The 2 faces are concave in the center and sometimes appear bent or slightly creased lengthwise. The attachment point, at the center of the base, is small, round, and slightly depressed. There is a very fine line around the circumference of the seed. Perianth remnants are often present, and there may be a style remnant.

Surface smooth; glossy. Under magnification it appears to have very fine slight wrinkles. Dark brown.

Length 2.3–2.5 mm; width 1.8–1.9 mm; thickness 0.7–0.8 mm.

Ditches; moist, disturbed soil; GP.

40. *Polygonum pensylvanicum* L.   Pennsylvania
smartweed

Achene: Outline nearly round with a small acuminate tip. Cross section a rounded oblong. Form discoid, with a narrow tip; often one face is somewhat concave or has a lengthwise crease. A style remnant may be present on the tip. The attachment point, located opposite the tip, is small and round and has a raised rim. With high magnification a faint lengthwise line may be visible on the center of each face.

Surface smooth; highly glossy. Under magnification it appears very finely wrinkled. Black to dark brown, with a very pale attachment point.

Length 2.8–3.1 mm; width 2.7–3.0 mm; thickness 0.9–1.0 mm.

Cultivated fields, roadsides; moist, disturbed soil; E 1/2 GP.

41. *Polygonum persicaria* L.   Lady's thumb

Achene (dimorphic, with a flat form and a three-sided form, both found on the same plant):

Flat form (more common): Outline short ovate with a small apical tip. Cross section a rounded oblong. Form ovoid, very compressed, thicker near the base. One face may have a distinct basal swelling. A small style remnant and perianth remnants may be present. The attachment point, located opposite the tip, is small and has a slightly raised rim.

Surface smooth; highly glossy. Black.

Length 2.0–2.2 mm; width 1.8–1.9 mm; thickness ±0.9 mm.

Three-sided form (less common): Outline short ovate with a small apical tip. Cross section 3-lobed. Form ovoid, with 3 prominent ridges alternating with 3 concave folds. A small style remnant and perianth remnants may be present. The attachment point, located opposite the tip, is small and has a slightly raised rim.

Surface smooth; highly glossy. Black.

Length 1.9–2.2 mm; thickness 1.4–1.8 mm; the ridges are ±0.6–0.7 mm thick.

Cultivated fields; moist disturbed soil; E 2/3 GP.

42. *Polygonum ramosissimum* Michx.   Bush knotweed

Achene: Outline broadly elliptical. Cross section triangular. Form 3-angled, with 3 elliptical, somewhat concave faces. The base is blunt and the apex acuminate. A style remnant and perianth remnants are usually present. The small round attachment point is at the base.

Surface smooth; very shiny, especially on the angles. At high magnification the faces appear slightly roughened. Black to brown-black.

Length 2.3–2.9 mm; width (of one face) 1.7–2.1 mm.

Pastures, rangeland, roadsides; damp soil; throughout GP.

43. *Polygonum scandens* L.   Climbing false buckwheat

Achene: Outline elliptical, with the widest part just below the middle. Cross section 3-pronged. Form ellipsoid, 3-winged. A style remnant and perianth remnants may be present. The attachment point is at the center of the base.

Surface smooth; highly glossy. Black; attachment point pale.

Length 3.3–3.6 mm; width (of one face) 1.8–2.2 mm.

Fence rows, waste places; open areas; SE 1/4 GP.

44. *Rumex acetosella* L.   Sheep sorrel

Achene (with or without adherent pericarp): Outline ovate with an extended base. Cross section triangular. Form 3-angled, with 3 broadly elliptical faces. The angles are rounded but well defined. The faces are flat or have a slight lengthwise crease. The apex is acute, and the extended base is blunt, with the attachment point at the center.

Surface usually covered by closely adherent perianth parts, making it dull and granular. There is a central lengthwise nerve on each of the 3 perianth parts. When the perianth is worn off the surface is smooth and glossy. Red-brown.

Length 1.3–1.5 mm; width (of one face) 0.9–1.0 mm.

Waste areas, lawns; acid soil; E central GP.

45. *Rumex altissimus* Wood   Pale dock

Achene: Outline broadly elliptical, tapering to an acute tip at the apex and with a short, truncate extension at the base. Cross section triangular with well-defined angles and straight or concave sides. Form 3-angled, with 3 broadly elliptical faces. Each face is slightly concave, especially near the ends, or has a faint lengthwise crease. The angles are narrowly winged. The attachment point is at the base.

Surface smooth; very shiny. Orange-brown.

Length 2.6–2.7 mm; width (of one face) 1.8–2.0 mm.

Waste places; moist soil; S 1/2 GP.

46. *Rumex crispus* L.   Curly dock

Achene: Outline ovate to broadly elliptical, with a narrow apex and short truncate base. Cross section triangular, with well-defined angles. Form 3-angled, with 3 elliptical faces. The angles are narrowly winged. The attachment point is at the base.

Surface smooth; very shiny. Dark orange brown.

Length 2.0–2.3 mm; width (of one face) 1.4–1.7 mm.

Clover and alfalfa fields, roadsides, waste areas; disturbed soil, open areas; throughout GP.

47. *Rumex mexicanus* Meisn.   Willow-leaved dock

Achene: Outline ovate, with an acute apex and slightly extended blunt base. Cross section triangular, with well-defined angles. Form 3-angled, with 3 elliptical faces. The faces are broadest below the center. The angles have distinct thin wings, less than 0.1 mm wide.

Surface smooth; shiny. Dark red-brown; the wings are often paler.

Length 1.9–2.2 mm; width (of one face) 1.1–1.5 mm.

Waste areas; sandy or saline soil, open areas; N 1/2 GP.

48. *Rumex patientia* L.   Patience dock

Achene: Outline ovate, with a narrow apex. Cross section triangular with very narrow angles. Form 3-angled, with 3 ovate faces. The angles are narrowly winged. At the basal juncture of the angles is a very short stalk.

Surface smooth; highly glossy. Light orange-brown; the wings are tinged with green, and the stalk is white.

Length 2.8–2.9 mm; width (of one face) 1.8–2.2 mm.
Waste areas; moist soil; central GP.

Clusiaceae
St. John's-wort
Family

49. *Hypericum perforatum* L.   Common St. John's-wort

Seed: Outline rounded oblong with a small tip at each end. Cross section round. Form cylindrical, with a small apical tip and a blunt basal tip that includes the hilum.

Surface pitted or roughened; dull. It appears shiny when viewed at 10× magnification. At high magnification it appears strongly sculpted in an irregular reticulate pattern. Light brown.

Length 1.0–1.2 mm; diameter 0.4–0.5 mm.
Overgrazed pastures, roadsides, waste areas; sandy soil; E central GP.

Malvaceae
Mallow Family

50. *Abutilon theophrasti* Medic.   Velvet-leaf, butterprint

Seed: Outline kidney shaped or rounded triangular with a deep notch. Cross section elliptical, constricted in the center. Reniform or ovoid, compressed, the 2 broad faces concave; a distinct notch along one edge divides the seed into 2 unequal lobes. One lobe is thinner and usually more angular than the other. The hilum, in the notch, is covered by a remnant of the funiculus, a long thin piece of tissue that is adherent to the longer, thinner lobe of the seed and is free and curved upward in the center of the notch.

Surface rough; very dull. When seen at 10× magnification it appears finely reticulate. Scattered white strigose hairs are visible at 8×; these are more abundant in the notch region. Gray-brown; the funicular remnant is greenish.

Length 2.9–3.4 mm; width 2.6–2.9 mm; thickness 1.4–1.6 mm.
Summer crops; open areas, fertile soils; E 1/3 GP.

51. *Hibiscus trionum* L.   Flower-of-an-hour,
                                               Venice mallow

Seed: Outline rounded triangular with a large shallow notch in one side. Cross section rounded oblong. Form ovoid, compressed, with a distinct notch in one edge that divides the seed into 2 unequal lobes. There is a depression in the center of each face. The ovate hilum is in the notch. The hilum is typically covered by a remnant of the funiculus, a long thin tissue that is adherent at the thinner end of the seed and is free and curved upward just above the center of the notch.

Surface rough; dull. At 10× magnification it appears finely reticulate and has scattered, sharply defined tubercles. At high magnification, these tubercles are transparent, yellow-brown, smooth, and shiny. Dark brown-black.

Length (along the notched edge) 2.0–2.2 mm; width 1.8–2.2 mm; thickness 1.3–1.5 mm.

Row crops, roadsides, waste places, gardens; fertile soils; E 1/2 GP.

52. *Malva neglecta* Wallr.   Common mallow

Seed: Outline kidney shaped to nearly round, with a small well-defined notch in the margin. Cross section oblong. Form like a narrow sector of a thick disk, with a notch in the inner angle. The seed is thicker above the notch (the upper end in fruit). The hilum is in the notch, usually covered with whitish tissue.

Surface finely textured; dull. At high magnification a very fine overall pattern of wrinkles is visible. Black with a grayish cast.

Length 1.6–1.8 mm; width 1.5–1.7 mm; thickness 0.9–1.2 mm.

Lawns, waste places, hay meadows; moist, fertile soil; throughout GP except NW.

53. *Malva rotundifolia* L.   Round-leaf mallow

Seed: Outline generally circular, with a deep, rounded notch, one side of which projects as a tooth. Cross section wedge shaped. Form like a sector of a sphere, with a deep, rounded notch in the inner angle; the flat faces slightly concave. The notch is off center, closer to the lower end in fruit; the hilum is in the notch.

Surface finely wrinkled; somewhat glaucous at 10× magnification. Dark brown (reddish when immature) with a thin whitish surface layer.

Length 1.7–1.9 mm; width (from notch to opposite side) 1.6–1.8 mm; thickness 1.0–1.2 mm.

Lawns, waste places, hay meadows; moist, fertile soil; throughout GP except SW.

54. *Malvastrum hispidum* (Pursh) Hochr.　　Yellow false-mallow

Seed: Outline ovate with a deep curved notch on one side. Cross section wedge shaped. Form sectorlike, narrow, with a deep notch located off center in the inner angle. The 2 broad faces are slightly sunken. Remnants of the funiculus are usually present as a bit of lacerate tissue at the edge of the notch near the smaller lobe of the seed and a thin flap of tissue attached to the inside of the notch.

Surface smooth; dull. At 10× magnification a network of very fine white nerves is visible. Dark red-brown to black; the notch is whitish.

Length 2.4–2.8 mm; width 2.1–2.5 mm; thickness 1.2–1.3 mm.

Waste areas, roadsides; dry soil, on gravel or limestone; SE 1/4 GP.

55. *Sida spinosa* L.　　Prickly sida

Seed: Outline with 1 straight side and 1 convex side that has a conspicuous deep notch near one end. Cross section wedge shaped, the angle less than 90°. Form sectorlike, with a conspicuous notch near one end of the curved outer face. The hilum is in the notch.

Surface smooth; dull. At 10× magnification it is very finely granular and has a soft sheen. Dark red-brown.

Length 2.1–2.2 mm; width (of one flat face) 1.3–1.5 mm.

Soybean fields, gardens; fertile soil; SE 1/4 GP.

**Violaceae
Violet Family**

56. *Viola rafinesquii* Greene　　Johnny-jump-up

Seed: Outline a rounded oblong with one end obliquely truncate and slightly wider. Cross section nearly round. Form ellipsoid.

Surface very smooth; lustrous. At high magnification it appears finely striate. Tan to straw colored.

Length 1.1–1.3 mm; thickness 0.6–0.7 mm.

Pastures; open areas; SE 1/4 GP

**Cucurbitaceae
Cucumber
Family**

57. *Cucurbita foetidissima* H.B.K.　　Buffalo-gourd

Seed: Outline pear shaped. Cross section a narrow ellipse. Form like a pear, very compressed, with rounded margins. A very narrow thin ridge is usually present around the rim. The hilum, ±1.0 mm long, extends from the small end of the seed down one side margin.

Surface nearly smooth; somewhat glossy. With magnification it appears finely roughened, but a smooth border, ±0.2 mm wide, surrounds each face. Light tannish yellow, with paler margins.

Length 7.1–9.0 mm; width 4.2–5.4 mm; thickness 1.6–2.1 mm.
Waste places, roadsides; sandy or gravelly soil; S 1/2 GP.

58. *Echinocystis lobata* (Michx.) T. & G. Wild cucumber

Seed: Outline elliptical with a broad notched extension at one small
end. Cross section a rounded oblong. Form ellipsoid, compressed, with
2 nearly flat faces; margins rounded. One end has an acute tip; the
other end has a thin, slightly spreading extension, ±3 mm wide, that is
divided into 2 teeth by a shallow notch. The hilum is in the notch.

Surface corky, with many large irregular pits and wrinkles; dull.
Light brown with a marbled pattern of dark brown. (The higher, flat
areas are light whereas the pits and depressed areas are dark.)

Length 1.6–2.0 cm; width 8.2–10.1 mm; thickness 2.8–4.0 mm.
Fence rows; moist rich soil, shaded areas; throughout GP except SW.

59. *Sicyos angulatus* L. Bur cucumber

Seed: Outline obovate with 2 distinct rounded teeth at the small end.
Cross section elliptical. Form obovoid, compressed, with 2 ovoid teeth
at the small end. The hilum is a straight narrow line along the margin
of the small end; the teeth point outward from the ends of the hilum.

Surface smooth; very dull. The seed appears to have a thin translucent coat over a dark surface. Dark brown; the teeth are paler.

Length 7.0–8.0 mm; width 5.2–6.0 mm; thickness 3.1–3.3 mm.
Fence rows, waste areas; damp soil; E 1/3 GP.

**Capparaceae
Caper Family**

60. *Cleome serrulata* Pursh Rocky Mountain bee plant,
spider flower

Seed: Outline ovate to teardrop shaped, with irregular margins. Cross
section short-elliptical. Form obovoid, compressed, apparently folded,
with the ends coming together to form the small end of the seed.
Form generally obovoid but with 1 or 2 flat areas near the broad end,
due to compression of the seed in the capsule. A conspicuous curved
groove extends from near the small end to the center of each face.
There is a thick lip on either side of each groove. The hilum, located
just above the small end, is small, round, and raised.

Surface appears coarsely tuberculate at low magnification. Dull but
with a slight sheen on smoother areas. At 10× magnification a dense
covering of fine superficial ridges is also visible. These ridges are
arranged in many sets of short parallel lines. Dark brown to black with
some patches of tannish ridges.

Length 3.2–3.7 mm; width 2.5–3.0 mm; thickness 1.8–2.2 mm.
Roadsides, pastures; disturbed soil, dry areas; throughout GP.

Brassicaseae
Mustard Family

61. *Barbarea vulgaris* R. Br.   Winter cress

Seed: Outline oblong with a notch in one short side. Cross section long ovate. Form somewhat oblong, compressed, with one long edge thinner and shorter than the other. The seed appears folded; on each face a groove extends lengthwise from the notch. (The cotyledons are accumbent, with the radicle forming the thinner and narrower side of the seed.) The hilum is in the notch, and a small bit of lacerate tissue fills the notch.

Surface finely textured; slight sheen. At high magnification a fine honeycomblike mesh is visible. Dull gray-brown; at high magnification there appears to be a glaucous yellowish or reddish coat over a dark red-brown surface.

Length 1.4–1.7 mm; width 1.1–1.3 mm; thickness 0.5–0.7 mm.
Winter-seeded small grains, pastures, roadsides, gardens; wet areas; E 1/3 GP.

62. *Berteroa incana* (L.) DC.   Hoary false alyssum

Seed: Outline round to slightly elongate. Cross section elliptical, ends narrow. Form generally ellipsoid. The embryo is evidently folded, with a curved groove on each face separating the narrow radicle from the cotyledons. The entire margin has a thin wing, ±0.1–.2 mm wide.

Surface finely roughened; slight sheen at 10×. At high magnification it is tuberculate. Dark red-brown; wing translucent, orange.

Length 1.3–1.8 mm; width 1.3–1.4 mm; thickness 0.6–0.7 mm.
Roadsides; disturbed soil; N 1/2 GP.

63. *Brassica juncea* (L.) Czern.   Indian mustard

Seed: Outline round or rounded oblong. Cross section round. Form slightly elongate, smoothly rounded. The indistinct hilum is near one small end of the seed, and a slight bit of white tissue projects from the hilum.

Surface smooth; dull. At 10× magnification a fine reticulate pattern is visible. Dark red-brown.

Length 1.4–1.9 mm; thickness 1.3–1.6 mm.
Grain fields, waste areas, roadsides; good soils; E 1/3 and N 1/4 GP.

64. *Brassica kaber* (DC.) Wheeler   Charlock, wild mustard

Seed: Outline round or broadly elliptical. Cross section round or ovate. Form globose or ellipsoid, slightly compressed. There is a short ridge of white tissue on one small end, covering the hilum.

Surface smooth; dull. When seen at 10× magnification there is an alveolate pattern over the entire surface. Dark orange-brown; the mesh is whitish.

Length 1.5–1.7 mm; width 1.4–1.6 mm; thickness 1.1–1.4 mm.
Grain fields, gardens; fertile soils; throughout GP except SW.

65. *Brassica nigra* (L.) Koch   Black mustard

Seed: Outline a broad ellipse. Cross section round. Form ellipsoid. At one small end is a very shallow notch that includes the hilum and a small ridge of irregular tissue.

Surface smooth; dull. At 10× magnification it appears shiny, and a fine but distinct reticulate pattern is visible. Dark red-brown; there are small black patches near the notch.

Length 1.3–1.6 mm; diameter 1.0–1.2 mm.
Grain fields, gardens; good soils; E central and N 1/2 GP.

66. *Camelina microcarpa* Andrz. ex DC.   Small-seeded false flax

Seed: Outline rounded oblong to ovate. Cross section rounded triangular, with indentations in the center of 1 or 2 sides. Form oblong, compressed. The seed appears folded with one half broader, thicker, and slightly longer than the other half; the shorter half ends in a narrow tip. (The incumbent cotyledons form the larger side, and the radicle the shorter side.) The seed is often somewhat angular because of compression in the fruit. There is a small ridge of white tissue at the hilum, on the small end of the seed.

Surface smooth; dull. At 10× magnification fine, evenly spaced tubercles can be seen, and the surface appears slightly glossy. Orange-brown; the small end may be darker.

Length 1.1–1.4 mm; width 0.5–0.9 mm; thickness 0.7–0.9 mm.
Cultivated fields, waste places; good soils; throughout GP.

67. *Camelina sativa* (L.) Crantz   Gold-of-pleasure

Seed: Outline oblong or elliptical. Cross section unequally 3-lobed. Form evidently folded (the cotyledons are incumbent). The radicle is narrow; its tip extends slightly beyond the cotyledons. There are deep grooves on either side of the radicle and shallow grooves between the cotyledons. There is a small bit of lacerate tissue in the attachment area between the radicle tip and the apex of the cotyledons.

Surface finely roughened; dull at 7× magnification. Orange-brown. Length 1.8–1.9 mm; diameter 0.9–1.1 mm.

Cultivated fields, waste places; good soils; N 1/3 GP.

68. *Capsella bursa-pastoris* (L.) Medic.   Shepherd's purse

Seed: Outline rounded oblong. Cross section ovate with wavy sides. Form ellipsoid, compressed. The seed is apparently folded; each broad face has 2 lengthwise curved grooves. (The cotyledons are incumbent; the deeper groove is between the radicle and the cotyledons, and the shallower groove separates the 2 cotyledons.) The basal end of the seed is thin and truncate, with a small thin piece of projecting tissue.

Surface smooth; dull. At 10× magnification it appears very finely textured and has a slight sheen. At high magnification it appears finely reticulate. Dull orange; the basal end is darker.

Length 0.9–1.0 mm; width 0.5 mm; thickness 0.3–0.4 mm.

Cultivated fields, lawns, gardens, waste places; moist to dry soil; E 3/4 GP.

69. *Cardaria draba* (L.) Desv.   Hoary cress, white-top

Seed: Outline ovate. Cross section ovate. Form ovoid, slightly compressed; one long edge is thinner and longer than the other. The seed is apparently folded; on each face there are 2 grooves, 1 distinct and 1 faint, that extend from the small end about 1/3 the length. (The cotyledons are incumbent, with the radicle forming the narrower edge of the seed. The deeper groove separates the radicle from the cotyledons, and the fainter groove separates the 2 cotyledons.) The hilum is in a notch at the small end of the seed.

Surface finely roughened; dull. At high magnification a fine honeycomblike mesh is visible, and the raised areas appear shiny. Orange-brown; the small end of the seed and the deeper groove are darker.

Length 1.8–2.1 mm; width 1.2–1.4 mm; thickness 0.8–0.9 mm.

Roadsides, waste places, cultivated fields; irrigated or alkaline soil; E central and N 1/4 GP.

70. *Chorispora tenella* (Pall.) DC.   Blue mustard

Seed: Outline a rounded short oblong. Cross section a narrow rounded oblong with indentations that divide it into 2 unequal segments. Form chiplike, bent, with 1 round thick end and 1 straight thin end. There is a deep groove on each face, extending from the straight end most of the length. (The cotyledons are accumbent; the groove separates the radicle from the cotyledons.) The radicle side makes up about 1/3 the width of the seed; it is thicker and slightly longer than the cotyledons, and it has a tapered tip. The cotyledon side is very thin toward the straight (basal) end of the seed. At the base there is a small piece of white tissue.

Surface finely roughened; dull. Dull yellow-orange.

Length 1.2–1.3 mm; width 0.8–0.9 mm; thickness ±0.2 mm.

Segments of the silique are often present and strongly adherent to the seeds.

Winter wheat fields, waste places; open areas, moist to dry soil; W 1/4 and central GP.

71. *Conringia orientalis* (L.) Dum.   Hare's-ear mustard

Seed: Outline rounded oblong. Cross section short ovate with 1 or 2 small notches. Form oblong, slightly compressed; evidently folded, with 1 distinct groove and sometimes a second shallow groove on each face. The cotyledons are usually incumbent. The radicle may be straight or curved; the radicle tip extends slightly beyond the cotyledons. There is a thin bit of lacerate tissue between the radicle and cotyledons in the basal attachment area.

Surface roughened or tuberculate; slight sheen when seen at 7× magnification. At 10× magnification it is distinctly tuberculate; the tubercles are low and closely spaced, arranged somewhat in lengthwise rows. Dark red-brown.

Length 2.1–2.6 mm; width 1.2–1.7 mm; thickness 1.1–1.4 mm.

Cultivated fields, especially grain fields; good soils; North Dakota and E central GP.

72. *Descurainia pinnata* (Walt.) Britt.   Tansy mustard
subsp. *brachycarpa* (Richard) Detling

Seed: Outline a rounded oblong. Cross section ovate or rounded triangular. Form oblong, compressed; apparently folded, with a lengthwise groove on each face. (The groove separates the radicle from the cotyledons.) The radicle side is thinner and tapers to an incurved tip at the small (basal) end of the seed. The cotyledon side is thicker and truncate at the basal end. The hilum is in the notch formed by the 2 parts; the notch is filled with a bit of translucent tissue.

Surface smooth; dull. At 10× magnification fine fingerprintlike ridges are visible. At high magnification these ridges are seen to be formed by many lengthwise rows of very small crosswise elliptical depressions. Light orange-brown, darker near the base.

Length 0.9–1.1 mm; width 0.5–0.6 mm; thickness 0.4–0.5 mm.
Grain fields, waste places; dry areas; throughout GP.

73. *Descurainia sophia* (L.) Webb　　　　　　Flixweed

Seed: Outline a rounded oblong. Cross section ovate to rounded triangular. Form oblong, rounded, with 1 convex and 1 nearly flat face. The seed is apparently folded, with 2 faint lengthwise grooves on each face. (The cotyledons are incumbent; one groove separates the radicle from the cotyledons, and the other groove is between the cotyledons.) The radicle side tapers to a narrow tip. The hilum is in the shallow notch between the radicle and cotyledons; the notch is filled by a thin piece of white tissue. Some seeds have a thin transparent marginal wing on the rounded end, opposite the notch.

Surface smooth; dull. At 10× magnification lengthwise striations are visible. At 25× these are seen to be formed of rows of minute depressions. Dull orange.

Length 0.9–1.3 mm; width 0.4–0.6 mm; thickness 0.3–0.5 mm.
Grain and alfalfa fields, waste places; dry areas; throughout GP.

74. *Erysimum cheiranthoides* L.　　　　　Wormseed wallflower

Seed: Outline variable; may be oblong, elliptical, or angular. Cross section variable, often plano-convex. Form apparently folded. (The cotyledons are incumbent.) The radicle may be straight or bent; its tip extends beyond the cotyledons. The folded end of the seed is often acute.

Surface finely roughened; slight sheen at 10× magnification. Light orange-brown.

Length 1.0–1.2 mm; width 0.5–0.6 mm; thickness 0.3–0.5 mm.
Cultivated fields; good soils; N 1/2 GP.

75. *Erysimum repandum* L.

Bushy wallflower, treacle mustard

Seed: Outline a rounded oblong. Cross section plano-convex. Form ellipsoid, with 1 flat face. The seed is apparently folded, with a lengthwise groove extending from the base most of the length of each face. (The groove marks the separation between the radicle and the cotyledons.) The radicle makes up about 1/3 the width of the seed and is slightly longer than the cotyledons. There is a membranous marginal wing, ±0.1–0.2 mm wide, at the rounded end of the seed and a similar wing on the cotyledon portion. The hilum is in the shallow notch between the radicle and cotyledons; the notch is filled with a thin piece of irregular tissue.

Surface finely roughened; slight sheen. Yellow-orange; the ends of the seed are darker, and the marginal wing is transparent.

Length 1.3–1.7 mm; width 0.6–0.7 mm; thickness 0.3–0.4 mm.

Cultivated fields, especially winter-seeded small grains, roadsides, waste places; open areas, often dry soil; S 1/2 GP.

76. *Lepidium campestre* (L.) R. Br.

Field peppergrass

Seed: Outline long ovate. Cross section rounded; there is 1 straight side and 1 strongly convex side. Form obovoid, with 1 flat face. (The cotyledons are incumbent; the outer side of the cotyledons forms the flat face of the seed, and the narrow radicle forms the convex face.) The radicle is outlined by distinct grooves that extend from the small end of the seed most of the length. At the small end there is an extension of translucent tissue, ±0.2 mm long.

Surface very finely wrinkled; dull. Dark red-brown.

Length 2.3–2.5 mm; width 1.0–1.2 mm; thickness 1.2–1.3 mm.

Clover, alfalfa, and winter wheat fields, waste areas; good soils; E central GP.

77. *Lepidium densiflorum* Schrad.

Peppergrass

Seed: Outline ovate, with 1 long straight side. Cross section narrowly triangular. Form obovoid, compressed; apparently folded, with a groove from the small end to the center of each broad face. (The cotyledons are accumbent.) The long straight edge of the seed (the cotyledon side) is thicker and has a narrow flat marginal face. Along the radicle side and around the wide end of the seed there is a conspicuous but very thin wing, ±0.1–0.2 mm wide. There is a small bit of tissue in the hilum area, at the small end.

Surface finely roughened; slightly glossy. At high magnification it appears finely tuberculate. Orange; the wing is darker but has a transparent edge.

Length 1.4–1.8 mm; width 0.9–1.2 mm; thickness 0.4–0.5 mm.
Roadsides, waste places; dry, compacted soil; throughout GP.

78. *Lepidium perfoliatum* L.   Clasping peppergrass

Seed: Outline obovate, either symmetrical or with 1 long straight side. Cross section oblong to narrowly ovate. Form obovoid, compressed; apparently folded, with a distinct curved groove from the small end to near the wide end of each face. A shallow lengthwise groove may also be present. (The cotyledons are incumbent; the deeper groove separates the cotyledons from the radicle, and the shallow groove lies between the cotyledons.) The cotyledon side of the seed is often thicker and straighter than the radicle side. A marginal wing, ±0.1 mm wide, may extend either around the entire margin or only around the radicle side and wide end.

Surface finely roughened; dull to glaucous. At 20× magnification it appears colliculate. Orange-brown; the wing is translucent pale yellow.

Length 1.8–2.0 mm; width 1.1–1.4 mm; thickness 0.5–0.6 mm.
Grain fields, waste areas; good soils; throughout GP.

79. *Lepidium virginicum* L.   Virginia peppergrass

Seed: Outline obovate, with 1 nearly straight long edge. Cross section narrowly triangular. Form compressed obovoid, asymmetric; apparently folded, with a distinct groove that extends from the small end to the center of each face, separating the radicle from the cotyledons. The radicle portion of the seed is slightly smaller than the cotyledon portion. The radicle tip is pointed. The cotyledon edge of the seed is thicker and rounded. There is a thin marginal wing around the radicle edge and broad end of the seed. The wing is ±0.1 mm wide. The attachment point is at the small end, between the radicle and cotyledons.

Surface finely roughened; slight sheen at 10× magnification. At 25×, the entire surface is uniformly tuberculate. Orange-brown; the wing is translucent, with an orange inner band and pale outer band.

Length 1.6–1.9 mm; width 1.1–1.2 mm; thickness 0.4–0.5 mm.
Crop fields, gardens; moist soil; SE 1/4 GP.

54

80. *Nasturtium officinale* R. Br.   **Watercress**

Seed: Outline elliptical or obovate. Cross section ovate. Form ellipsoid, compressed; apparently folded, with a faint groove on each face. (The cotyledons are accumbent.) The wider, thinner side of the seed is formed by the cotyledons; the radicle is thicker but narrow and tapered to a small tip. A shallow notch at the base is filled with a bit of thin tissue.

Surface distinctly reticulate; shiny. Orange-brown.

Length 0.9–1.0 mm; width 0.7–0.8 mm; thickness 0.3–0.4 mm.

Low pastures and roadsides; wet places; S 2/3 GP.

81. *Rorippa palustris* (L.) Bess.   **Bog yellow cress**

Seed: Outline variable; generally ovate with a notch in the wide end. Cross section ovate. Form usually ovoid, compressed but of irregular thickness. The seed is apparently folded; there is a broad lengthwise groove on each face. (The cotyledons are accumbent.) The radicle is tapered to a small tip, and the radicle side is thinner and narrower than the cotyledon side. The notch at the wide (basal) end of the seed is filled with thin whitish tissue.

Surface finely textured; appears glittery when seen at 10× magnification. At high magnification it appears distinctly colliculate and shiny. Yellow-brown; the notch area is darker.

Length 0.6–0.7 mm; width 0.6 mm; thickness 0.3 mm.

Roadside ditches, low pastures; wet places; throughout GP except SE 1/4.

82. *Rorippa sinuata* (Nutt.) Hitchc.   **Spreading yellow cress**

Seed: Outline variable, generally angular, 3- or 4-sided. Cross section variable, often 3- or 4-angled. Form generally oblong and angular. One face is convex, the others slightly concave, so that the angles between the faces appear ridged. The indistinct hilum is located in a notch at one end of the seed.

Surface finely textured and somewhat shiny at 10× magnification. At 25× magnification it appears colliculate. Light brown; the hilum area is darker.

Length 0.7–0.9 mm; width 0.6–0.8 mm; thickness 0.4–0.6 mm.

Roadside ditches; moist soil; throughout GP except NE 1/8.

83. *Sisymbrium altissimum* L.   **Tumbling mustard**

Seed: Outline a rounded oblong. Cross section ovate with a notch in each long edge. Form oblong, compressed; evidently folded, with a distinct groove on each broad face. (The cotyledons are incumbent.) The radicle makes up about 1/3 the width of the seed. The broad faces of the seed may be similarly rounded, or one of them may be flat and the other conspicuously humped in the center. At the base there is a small notch that is filled with a thin bit of tissue.

Surface smooth; dull. At high magnification it appears very finely tuberculate, semiglossy. Dull orange.

Length 1.1–1.3 mm; width 0.6–0.8 mm; thickness 0.4–0.6 mm.

Grain fields, cultivated fields, gardens, roadsides; good soils; throughout GP.

84. *Sisymbrium loeselii* L.   **Tall hedge mustard**

Seed: Outline variable; may be rounded oblong, elliptical, or asymmetrically 3- or 4- angled. Cross section variable, often 3-sided, sometimes with shallow notches. Form generally oblong, slightly compressed; evidently folded, with a distinct curved groove on each face. The cotyledons are incumbent. The radicle makes up about 1/3 the width of the seed; it may be straight or bent toward one side. The radicle tip extends slightly beyond the cotyledons, and there is a bit of lacerate tissue in the attachment area, between the radicle tip and cotyledons.

Surface finely roughened; shiny at 10× magnification. At 30×, it appears finely tuberculate. Orange.

Length 0.7–1.0 mm; width 0.5–0.6 mm; thickness 0.3–0.4 mm.

Cultivated fields, roadsides; good soils; N 1/2 GP.

85. *Sisymbrium officinale* (L.) Scop.   **Hedge mustard**

Seed: Outline variable; elliptical to rounded oblong or with 1 straight edge and 1 convex edge. Cross section variable; often a rounded V-shape. Form oblong, compressed; evidently folded, with 2 lengthwise grooves on each face. (The cotyledons are incumbent.) The radicle is narrow, tapered to a small tip, and often curved toward one face; its position relative to the cotyledons varies. There is a notch between the radicle and cotyledons at the base of the seed; the notch is filled with white tissue.

Surface dull; finely roughened. At high magnification it appears irregularly tuberculate and semiglossy. Orange-brown.

Length 1.1–1.6 mm; width 0.5–0.7 mm; thickness 0.3–0.5 mm. *Cultivated fields, waste places; good soils; E 1/3 GP.*

86. *Thlaspi arvense* L.   Pennycress, fan weed

Seed: Outline generally ovate; one long edge has a more pronounced curve than the other. Cross section long ovate. Form ovoid, compressed; apparently folded, with a groove extending from the small end to about the center of each face. (The cotyledons are accumbent.) The radicle side forms the straight edge and is thinner than the cotyledon side. A bit of white tissue extends ±0.1 mm from the small (basal) end of the seed.

Surface covered by ridges arranged in concentric loops. These are barely visible without magnification. There are ±6 loops on each broad face. The ridges are somewhat wavy in form and uneven in height. With magnification the surface appears glittery or shiny. Red-brown; the ridges and basal area are slightly darker.

Length 1.7–1.9 mm; width 1.1–1.2 mm; thickness 0.6–0.7 mm.

Waste places, winter cereals, especially winter wheat, roadsides, gardens, pastures; good soils in open areas; GP except SW 1/8.

Mimosaceae Mimosa Family   ### 87. *Desmanthus illinoensis* (Michx.) MacM.   Illinois bundle-flower

Seed: Outline variable; may be rounded triangular, rhombic, or elliptical. Cross section rounded oblong. Form more or less ovoid, compressed, of irregular thickness, with 2 to 4 rounded corners. One corner is extended to form a distinct tip. The whole seed may be somewhat bent. The hilum is located along one margin, just below the tip. The hilum is an elliptical raised area, ±2.0 mm long and 1.0 mm wide. On each face of the seed is a narrow but distinct V-shaped line that opens toward the tip. The line is just visible without magnification and is conspicuous with low magnification.

Surface smooth; somewhat glossy. Orange-brown; the V-shaped line is darker, and the hilum is blackish.

Length 3.3–3.8 mm; width 2.3–2.6 mm; thickness 1.0–1.1 mm.

Roadsides, pastures; rocky soil; S 2/3 GP.

Caesalpinaceae Caesalpinia Family   ### 88. *Cassia chamaecrista* L.   Showy partridge pea

Seed: Outline rhombic, with 2 nearly straight sides opposite 2 curved sides and with a tip projecting from one corner. Cross section ellipti-

57

cal. Form rhomboid, compressed, with rounded margins and 1 extended corner. On each broad face a slight rounded ridge extends from the tip toward the center. The hilum is located on the curved margin below the tip. The hilum is dark, round, and small. Just above the hilum is a very small, downwardly curved hook of tissue.

Surface smooth, with a waxy sheen. Faint striations are visible without magnification. When viewed at 7×, there are irregularly spaced lengthwise rows of very shallow pits. Each of these circular pits has a central dotlike depression. Black-brown; the margins are yellowish, reddish, or gray.

Length 3.4–4.0 mm; width 2.3–2.9 mm; thickness 1.2–1.3 mm. (Length measured from the tip along the adjacent straight side.)

Roadsides; sandy or rocky soil; E central and S 1/3 GP.

89. *Hoffmanseggia glauca* (Ort.) Eifert   Indian rush-pea

Seed: Outline ovate, with the small end bent slightly to one side. Cross section oblong. Form ovoid, compressed, with rounded margins. The small ovate hilum is on the margin, just below the bent small end.

Surface smooth; glossy. When seen at high magnification it appears finely textured. Black.

Length 3.7–4.2 mm; width 2.6–2.8 mm; thickness 0.8–1.0 mm.

Roadsides, cultivated fields; rocky or sandy soil; SW 1/4 GP.

Fabaceae
Bean Family

90. *Amorpha canescens* Pursh   Lead plant

Seed: Outline elliptical, with one end curved to form an S-shaped tip. Cross section ovate. Form elongate, rounded, compressed, with one end curved and tapered to a narrow tip. Just below the tip is a notch that includes the ovate hilum. There is a dark ring around the hilum, and a dark line extends along the margin from the hilum to the far end of the seed.

Surface very smooth; shiny. Light red-brown.

Length 2.1–2.5 mm; width 1.1–1.2 mm; thickness 0.8–1.0 mm.

Upland pastures, roadsides; dry soil; throughout GP.

91. *Astragalus canadensis* L.   Canada milk-vetch

Seed: Outline more or less reniform, with a small notch between 2 lobes. Cross section oblong. Generally reniform, lobes unequal; compressed, nearly flat. The hilum is in the notch, on the side nearer the small lobe; it is small, round, with a whitish ring around a small depression.

Surface smooth; dull. When seen at 10× magnification it has a waxy sheen. Light greenish brown to light orange-brown; hilum dark.

Length 1.7–2.2 mm; width 1.1–1.4 mm; thickness 0.6–0.8 mm.

Stream banks, thickets, roadsides; moist soil; throughout GP except SW 1/4.

92. *Astragalus mollissimus* Torr.   **Woolly locoweed**

Seed: Outline variable; may be reniform, rounded triangular with a notch in one side, or oblong. Cross section ovate. Generally reniform, with one lobe thicker than the other, but sometimes wider than long (width measured from the hilum to the opposite side). The 2 broad faces may be rounded or flat. The hilum, in the notch, is a small round depression.

Surface smooth; dull. Under magnification there is a soft sheen. Olive, orange-brown, or purplish brown; the hilum area is paler, and at high magnification a uniform pattern of fine dark dots is visible.

Length 1.6–2.3 mm; width (perpendicular to the hilum) 1.5–2.2 mm; thickness 0.9–1.4 mm.

Rangeland, pastures; dry areas; SW 1/4 GP.

93. *Coronilla varia* L.   **Crown vetch**

Seed: Outline rounded oblong with a shallow notch in the center of one long side. Cross section elliptical. Form cylindrical, slightly compressed, with rounded ends and a shallow rounded notch near the center of one long edge. The very small round hilum is located in the notch.

Surface smooth; glossy. Deep purplish red.

Length 3.1–3.7 mm; width 1.0–1.3 mm; thickness 0.9–1.1 mm.

Roadsides, banks, usually spreading from plantings; open areas; E central GP.

94. *Desmodium illinoense* A. Gray   **Illinois tickclover**

Seed: Outline ovate, with a shallow notch in the center of one long side. Cross section ovate or elliptical. Form ovoid, compressed, both faces slightly depressed in the center. One long margin is slightly thicker than the other. The thicker margin has a small notch that includes the hilum. The hilum is ±0.5 mm in diameter and has a distinct collar.

Surface smooth; glossy. Dull yellowish or greenish brown; the collar is darker.

Length 3.4–3.7 mm; width 2.5–2.6 mm; thickness 1.2–1.4 mm.
Roadsides; moist soil; SE 1/4 GP.

95. *Lespedeza cuneata* (Dumont) G. Don   Sericea lespedeza, Chinese bush clover

Seed: Outline ovate, with a shallow notch near the wide end. Cross section ovate. Form ovoid, compressed, thinner toward the notched margin. The hilum, in the notch, is surrounded by a rim of translucent tissue. The diameter of the hilum with its rim is ±0.2 mm. On each face there may be a faint groove extending from the notch toward the other end of the seed.

Surface smooth; somewhat glossy. Olive-brown or greenish; the area around the hilum is tinged red.

Length (1.3) 1.6–1.8 mm; width 1.0–1.3 mm; thickness 0.6–0.8 mm.
Roadsides, pastures; open areas; SE 1/8 GP.

96. *Lespedeza stipulacea* Maxim.   Korean lespedeza, Korean clover

Seed: Outline broadly elliptical, with a shallow notch near one end. Cross section narrowly ovate. Form ellipsoid, compressed, thinner toward the notched long margin. The hilum, in the notch, is ±0.2 mm in diameter and has a distinct collar. On each face there may be a slight groove extending from the notch, parallel to the edge of the seed. The 2 faces may be flat or depressed.

Surface smooth; glossy. Dark red-brown to black; the hilum and a small area around it are pale with a reddish or greenish tinge.

Length 1.7–2.2 mm; width 1.4–1.7 mm; thickness 0.8–0.9 mm.
Roadsides, lawns, pastures; open areas; SE 1/4 GP.

97. *Lotus corniculatus* L.   Bird's-foot trefoil

Seed: Outline somewhat ovate, with a small notch in one long side. Cross section ovate. Form ovoid, slightly compressed, thinner toward the notched margin. The hilum, in the notch, is round, ±0.2 mm in diameter.

Surface smooth; dull. Olive, dull brown, or dark brown, sometimes with small dark mottles. There is a small dark spot above the hilum and a pale area surrounding the hilum.

Length 1.3–1.6 mm; width 1.0–1.3 mm; thickness 0.8–1.2 mm.
Roadsides, lawns; good soils; SE 1/4 GP.

98. *Medicago lupulina* L.   Black medick, nonesuch

Seed: Outline ovate to somewhat reniform. Cross section ovate. Form ovoid, compressed, with a small notch on one long margin about 1/3 the distance from the small end. The lower edge of the notch protrudes, and a slight linear depression may extend diagonally inward from the notch across each face. The small round hilum is in the notch.

Surface smooth; dull. Yellow-brown with green and orange tinges; the hilum is darker. Below the hilum there is a light-colored V-shaped area, and there are small dark patches above and below the V. A pale streak extends across each face from the hilum toward the wide end of the seed.

Length 1.2–1.6 mm; width 0.9–1.2 mm; thickness 0.6–0.9 mm.
Roadsides, lawns, pastures; moist or dry open areas; throughout GP.

99. *Medicago minima* (L.) Bartal.   Small bur-clover

Seed: Outline kidney shaped. Cross section ovate. Reniform with a distinct curved notch in the inner margin near one end. The small round hilum is in the notch. Next to the hilum (closer to the center) the chalaza forms a conspicuous rounded extension of the margin. From the notch a slight groove runs across each face toward the far end of the seed.

Surface smooth; dull. Light orange-yellow; the chalaza and notch area are darker, and the grooves are pale.

Length 1.8–2.2 mm; width 0.9–1.0 mm; thickness 0.6–0.7 mm.
Roadsides, waste places; open areas; S 1/4 GP.

100. *Medicago sativa* L.   Alfalfa

Seed: Outline varies from somewhat reniform to ovate with a notch on one long side near the wide end. Cross section ovate. Ovoid to reniform, compressed. The small round hilum is in the notch. A slight groove may extend from the hilum diagonally across each face.

Surface smooth; dull. With high magnification it appears finely textured. Dull orange-brown or olive; the area around the hilum is slightly darker.

Length 2.0–2.5 mm; width 1.3–1.6 mm; thickness 0.9–1.1 mm.
Roadsides, waste areas, escaping from cultivation; throughout GP.

101. *Melilotus alba* Medic.   White sweet clover

Seed: Outline ovate, with a distinct large notch on one long side near the broad end. Cross section ovate. Ovoid, slightly compressed, thinner toward the notched margin; the 2 faces are usually depressed. The small round hilum is in the notch. On each face a shallow groove extends from the hilum 2/3 the distance to the small end of the seed.

Surface smooth; dull. Yellow-brown, greenish, or orangish. Above the hilum is a small dark spot; below the hilum is a dark V-shaped area. The grooves are pale.

Length 1.8–2.0 mm; width 1.3–1.5 mm; thickness 1.0–1.1 mm.
Roadsides, waste places; open areas, also cultivated; throughout GP.

102. *Melilotus officinalis* (L.) Pall.   Yellow sweet clover

Seed: Outline elliptical to ovate with a small notch near the end of one long side. Cross section ovate. Form ellipsoid to ovoid, compressed, thinner toward the notched margin; the 2 faces may be slightly depressed. The round hilum is in the notch. A slight groove extends from the hilum diagonally across each face.

Surface smooth; dull. Dull orange-yellow to buff, sometimes with a green tinge; some seeds have small brown mottles. There is a small dark spot just below the hilum. Pale streaks may extend diagonally from the hilum across each face.

Length 1.7–1.9 mm; width 1.0–1.3 mm; thickness 0.9–1.0 mm.
Roadsides, waste places, also cultivated; open areas; throughout GP.

103. *Oxytropis lambertii* Pursh   Purple locoweed

Seed: Outline variable but somewhat reniform with 2 unequal lobes. Cross section ovate. Form rounded to compressed, generally ovoid with a deep asymmetrical notch in one long margin. The very small round hilum is in the notch.

Surface smooth; semiglossy. Brown; paler in the notch area.
Length 1.8–2.2 mm; width 1.5–2.1 mm; thickness 0.7–1.0 mm.
Rangeland, pastures; dry open areas; throughout GP.

104. *Strophostyles helvola* (L.) Ell.   Wild bean

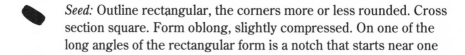

Seed: Outline rectangular, the corners more or less rounded. Cross section square. Form oblong, slightly compressed. On one of the long angles of the rectangular form is a notch that starts near one end of the seed and extends about 3/4 the length of the seed. The conspicuous elliptical hilum is in the notch. The hilum is a corky white area, ±4–5 mm long and 1.2 mm wide, surrounded by a smooth, shiny rim.

Surface smooth, hard, shiny, but more or less covered with dull, soft scurfy tissue. This scurfy material can be rubbed off. The hard surface is dark brown; with magnification it is pale brown with small brown-black mottles. When the scurfy layer is present the seed appears dark brown or sometimes paler.

Length 6.4–7.8 mm; width (from hilum to opposite angle) 3.9–4.1 mm; thickness 3.5–4.0 mm.

Fence rows, roadsides; rocky or sandy soil; SE 1/4 GP.

105. *Strophostyles leiosperma* (T. & G.) Piper — Slick seed bean

Seed: Outline rhombic with rounded corners, a rounded oblong, or an oblong with one round end and one angular end. Cross section nearly round. Form cylindrical or cylindrical with rounded ends; slightly compressed. The conspicuous hilum is located along the length of the side. It is elliptical, ±1.5–2.0 long and 1 mm wide, with a very dark, slightly raised rim around a very pale, distinctly raised corky center. At one end of the hilum is a raised knob, ±0.4 mm in diameter.

Surface smooth; somewhat shiny. Gray to buff with many fine, small black mottles.

Length 3.0–4.2 mm; width (from hilum to opposite edge) 2.7–3.0 mm; thickness 2.1–2.7 mm.

Fence rows, cultivated fields, roadsides; sandy or rocky soil; S 2/3 GP.

106. *Trifolium campestre* Schreb. — Low hop-clover

Seed: Outline ovate with a slight notch on one side near the small end. Cross section elliptical. Form ovoid, slightly compressed, with a shallow notch in one long margin. The small indistinct hilum is located in the notch. A slight groove runs from the hilum diagonally toward the center of each face.

Surface smooth, with a waxy sheen. Orange-yellow; a few seeds have orange patches. There is a dark V-shaped mark below the hilum, and the grooves are pale.

Length 1.0–1.2 mm; width 0.7–0.8 mm; thickness 0.6–0.7 mm.

Roadsides, waste places, pastures; rocky or sandy soil; E central GP.

107. *Trifolium pratense* L.   **Red clover**

Seed: Outline ovate with a large notch in one side near the small end. Cross section ovate. Form ovoid, compressed, with a notch in one long margin. The hilum, in the notch, is ±0.2 mm in diameter. On each face a slight groove extends from the notch toward the wide end of the seed.

Surface smooth; slightly glossy or dull. Straw colored, with or without purple mottles, to solid dull purple.

Length 1.9–2.3 mm; width 1.4–2.0 mm; thickness 1.0–1.1 mm.

Roadsides, pastures, cultivated fields, waste areas; moist soil; throughout GP except SW 1/8 and NW 1/8.

108. *Trifolium repens* L.   **White clover**

Seed: Outline variable; generally rounded with a broad shallow notch. Cross section ovate. Form ovoid, somewhat compressed, thinner toward the notched margin. The seed appears bent into a V-shape, with a notch and a groove between the ends. The small round hilum is in the notch.

Surface smooth; dull. When viewed at 10× magnification there is a slight sheen. Dull yellow or orange-brown. Below the hilum is a dark spot. A pale streak extends from the notch across each face.

Length 0.8–1.1 mm; width 0.8–1.0 mm; thickness 0.5–0.7 mm.

Lawns, pastures; moist soil; throughout GP except SW 1/8.

109. *Vicia villosa* Roth   **Hairy vetch**

Seed: Outline round. Cross section a broad ellipse. Form nearly globose, slightly compressed. The hilum is on a slightly elevated area along the margin between the 2 faces. The hilum is oblong, ±1.5–2.0 mm long and 0.5–0.7 mm wide, and has a narrow central line. The hilum is level with the surface except for a small groove around its edge.

Surface smooth; dull, with a fine velvety sheen. Brown-black; under magnification there is dark brown mottling on a brown or greenish brown background.

Diameter 3.8–5.1 mm; thickness 3.3–4.3 mm.

Roadsides, cultivated fields; persisting from cultivation on sandy soil; central and SE 1/4 GP.

Onagraceae
Evening Primrose
Family

110. *Gaura longiflora* Spach　　　　　Large-flowered gaura

Dry accessory fruit: Outline long obovate. Cross section round or nearly round with 4 convex sides. Form obovoid, elongate, with 4 faces. Each face has a central lengthwise ridge. The basal attachment area is round, hollow, oblique. At the apex, the angles between the faces form 4 small teeth.

Surface slightly roughened and wrinkled; dull, papery sheen. Tan with green and red-violet tinges.

Length 5.7–7.5 mm; diameter 1.8–2.0 mm.

Waste areas, roadsides; rocky slopes; SE 1/4 GP.

111. *Gaura parviflora* Doug.　　　　　Velvety gaura

Dry accessory fruit: Outline long obovate with the wide end extended into a narrow tip. Cross section with 4 convex sides, each with a small tooth in the center. Form obovoid, elongate, somewhat 4-sided with 4 rounded angles. The 4 faces meet in an acute apical tip. Near the tip the faces are depressed, making the angles more prominent. There is a central lengthwise ridge on each face. One side of the fruit is obliquely truncate at the base. The attachment point, in the truncate area, has a round rim and a hollow center.

Surface papery; dull, finely wrinkled. Light to medium tan, sometimes tinged with green.

Length 6.0–6.6 mm; diameter 1.7–2.4 mm.

Roadsides, waste areas, old fields; rocky slopes; S 1/2 GP.

112. *Oenothera biennis* L.　　　　Common evening primrose

Seed: Outline variable, usually angular; may be oblong, trapezoidal, or triangular. Cross section variable, usually angular with 3 or 4 unequal sides. Form like a narrow truncate pyramid, a tetrahedron, or a form with 3 flat faces and 1 long curved face; in any case the angles between the faces are very narrowly winged.

Surface finely roughened; dull. There are several fine lengthwise wrinkles on each face as well as irregular wrinkles near the angles. Dark orange-brown.

Length 1.0–1.7 mm; width 0.6–1.4 mm; thickness 0.5–1.0 mm.

Roadsides, waste places; moist soil; central and E central GP.

113. *Oenothera laciniata* Hill   Cut-leaved evening primrose

Seed: Outline irregular and angular, somewhat oblong or rhombic. Cross section variable, usually ovate or triangular. Form various, with both rounded and flat faces.

Surface alveolate; dull. Orange-brown.

Length 0.9–1.3 mm; thickness 0.7–1.0 mm.

Fallow fields, roadsides, waste places; open areas, dry soil; SE 1/4 GP.

Euphorbiaceae Spurge Family

114. *Acalypha monococca* (Engelm.) P. Mill.   Slender copperleaf

Seed: Outline obovate with an acute small end. Cross section round or slightly compressed. Form ovoid, slightly compressed; the dorsal face is distinctly convex, and the ventral face is convex but divided into 2 parts by the lengthwise raphe. Near the small end the 2 faces meet in a marginal rim. The small end of the ventral face (the chalazal region) is depressed and has a conspicious caruncle of translucent yellow tissue. The caruncle is a ridge from the tip about 1/3 the length of the seed. The raphe is a narrow line, extending from the caruncle to the center of the broad end of the seed.

Surface with many wrinkled ridges, oriented lengthwise. These ridges are clearly visible with low magnification. Mature seeds appear to have a dark red coat, overlaid with a thin glaucous grayish white layer, so that the surface appears gray-brown and waxy. Less mature seeds may lack the caruncle; they are pale green with brown mottlling; dull.

Length 1.7–1.8 mm; width 1.3–1.4 mm; thickness 1.1–1.3 mm.

Pastures, roadsides; rocky soil; SE 1/4 GP.

115. *Acalypha ostryaefolia* Ridd.   Hop-hornbeam copperleaf

Seed: Outline obovate with an acute to acuminate small end. Cross section nearly round. Form obovoid. On the ventral side a pale yellow translucent caruncle extends about 1/3 the length from the small end. The raphe is visible as a narrow dark line from the caruncle to the wide end.

Surface tuberculate with the tubercles arranged somewhat in lengthwise rows; glaucous. Gray; under magnification there are reddish undertones.

Length 2.0–2.2 mm; width 1.5–1.7 mm; thickness 1.4–1.6 mm.

Summer crop fields, gardens, waste places; moist or shaded areas; SE 1/4 GP.

116. *Acalypha rhomboidea* Raf.   Rhombic copperleaf

Seed: Outline obovate with a fine tip at the small end. Cross section nearly round or compressed so that one side is more convex than the other. Form ovoid, somewhat compressed; the dorsal face is slightly more convex and has a small tip that is curved over the end of the seed toward the flatter ventral face. On the ventral face there is a conspicuous narrow caruncle of yellowish translucent tissue that forms a ridge from the tip about 1/3 the length of the seed. The narrow raphe runs from the caruncle to the center of the broad end of the seed.

Surface smooth; slightly glossy. With magnification a fine scalariform pattern can be seen. On the dorsal face this pattern is oriented lengthwise; on the ventral face it has no definite orientation. Dark gray to black overall; with magnification, individual seeds are seen to vary from gray-brown with black mottling to mostly black.

Length 1.7–1.8 mm; width 1.0–1.3 mm; thickness 1.0–1.1 mm.
Cultivated fields, waste places; moist soil; E central GP.

117. *Acalypha virginica* L.   Virginia copperleaf, three-seeded mercury

Seed: Outline obovate with a pointed tip. Cross section round or slightly compressed. Form like a slightly compressed teardrop. On the ventral face a conspicuous caruncle of translucent yellowish tissue extends from the tip 1/3 to 1/2 the length of the seed. The narrow raphe runs from the caruncle to the center of the broad end of the seed.

Surface finely roughened; dull. With magnification, the entire surface is seen to be covered with irregular round tubercles and has a waxy sheen. There appears to be a dark red-brown undercoat covered with a translucent grayish layer, so that the tubercles are dark and the low areas grayish; individual seeds vary from predominantly brown to solid gray.

Length 1.8–2.1 mm; width 1.5–1.7 mm; thickness 1.3–1.5 mm.
Roadsides, waste areas; moist or dry soil; SE 1/4 GP.

118. *Croton capitatus* Michx.   Woolly croton

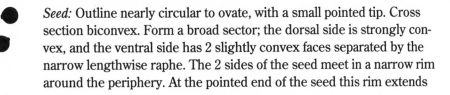

Seed: Outline nearly circular to ovate, with a small pointed tip. Cross section biconvex. Form a broad sector; the dorsal side is strongly convex, and the ventral side has 2 slightly convex faces separated by the narrow lengthwise raphe. The 2 sides of the seed meet in a narrow rim around the periphery. At the pointed end of the seed this rim extends over toward the ventral side. On the ventral side is a prominent fan-shaped caruncle, with its broad end at the tip of the seed. Just beyond the point of the caruncle is the small hilum, with a bit of ragged tissue attached.

Surface smooth; glossy. At 10× magnification fine wrinkles are visible. Tan, mottled with dark brown; the rim and the raphe are lighter, and the caruncle is very pale.

Length 4.7–5.0 mm; width 4.6–4.8 mm; thickness 2.8–3.2 mm.
Pastures, roadsides; rocky or calcareous soil; SE 1/4 GP.

119. *Croton glandulosus* L. var. *septentrionalis* Muell. Arg.   Tropic croton

Seed: Outline ovate; the small end has a pointed tip, and the persistent broad caruncle extends beyond the small end. Cross section biconvex. Form ovoid, compressed; the dorsal side is convex, and the ventral side has 2 slightly convex faces meeting at an obtuse angle. The raphe is a distinct line on this angle. The caruncle is 2-lobed, broader than long, located on a flat area of the ventral side near the small end of the seed. The hilum is small, indistinct; located beneath the notch of the caruncle.

Surface with a faint scalariform pattern; very shiny. Tan, with brown and dark brown mottles; raphe pale. Caruncle orange, with a waxy sheen.

Length (excluding caruncle) 3.3–3.7 mm; width 2.1–2.6 mm; thickness 1.6 mm.
Pastures, roadsides, waste places; sandy soil; SE 1/4 GP.

120. *Croton monanthogynus* Michx.   One-seeded croton

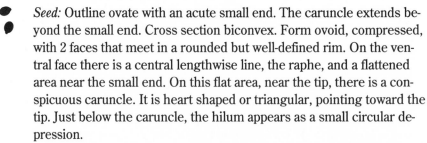

Seed: Outline ovate with an acute small end. The caruncle extends beyond the small end. Cross section biconvex. Form ovoid, compressed, with 2 faces that meet in a rounded but well-defined rim. On the ventral face there is a central lengthwise line, the raphe, and a flattened area near the small end. On this flat area, near the tip, there is a conspicuous caruncle. It is heart shaped or triangular, pointing toward the tip. Just below the caruncle, the hilum appears as a small circular depression.

Surface smooth; glossy. With magnification a fine alveolate texture is visible. Dark gray-brown; there appears to be an outer hard glaucous coat that is white or tan. The raphe is pale; the caruncle is pale and translucent.

Length 3.6–3.7 mm; width 2.6–2.8 mm; thickness 1.9–2.0 mm.
Pastures, roadsides, waste places; calcareous soil; SE 1/4 GP.

121. *Croton texensis* (Kl.) Muell. Arg.   Texas croton

Seed: Outline broadly ovate with a short, blunt tip. Cross section biconvex. Form like a slightly compressed sphere; the dorsal side is hemispherical, whereas the ventral side consists of 2 convex faces meeting at an obtuse angle. The angle is marked with a narrow line, the raphe. The ventral and dorsal sides meet in a narrow rim. On the ventral side, just below the tip, is a prominent caruncle that is 3-sided or heart shaped, ±1.1 mm wide. The indistinct hilum is below the caruncle.

Surface smooth; somewhat lustrous. Solid gray-brown or streaked and mottled with buff, gray-brown, light brown, and dark brown. The raphe is pale; the caruncle is translucent pale yellow.

Length 4.0–4.7 mm; width 3.5–4.0 mm; thickness 2.4–2.8 mm.
Pastures, rangeland, roadsides; sandy soil; S central and W central GP.

122. *Euphorbia corollata* L.   Flowering spurge

Seed: Outline oblong to obovate, with a small blunt tip. Cross section nearly round. Form ovoid, nearly the same width for most of its length. The boundary between the dorsal and ventral faces is marked by a slight ridge. The dorsal face is convex. The ventral face is divided in half by a lengthwise line, the raphe. The wide end of the seed is slightly flattened. On the ventral face, near the small end, is a round depressed area, the chalaza. The hilum is in the depressed area, and there is a small caruncle at the edge of the hilum. The raphe runs from the hilum to the center of the broad end of the seed.

Surface with a thick, dull, granular yellowish white coating that may appear distinctly punctate when seen with magnification. The raphe is dark. The coating becomes thinner with maturity and may be worn off. Without it the seed is finely textured and has a slight sheen. Black when the outer coat is absent.

Length 2.7–2.9 mm; width 1.9–2.1 mm; width 1.8–2.0 mm.
Roadsides; dry open soil; E central and SE 1/8 GP.

123. *Euphorbia cyathophora* Murray   Fire-on-the-mountain

Seed: Outline ovate; one end is straight, and the other end has an abrupt narrow tip. Cross section round. Form ovoid to nearly globose, with a flat end. The hilum is indistinct, located in the center of a flat

69

area near the small end of the seed. The flat area is ±1 mm in diameter. From the hilum a narrow line extends to the center of the flat end of the seed.

Surface very rough with many short, rather sharp projections of varying size; dull. Under high magnification the hilum area appears finely crystalline. Medium to dark brown or brownish black; the projections are paler or orangish.

Length 2.6–2.9 mm; thickness 2.4–2.6 mm.

Waste areas; moist soil; E central GP.

124. *Euphorbia dentata* Michx.   Toothed spurge

Seed: Outline short ovate, flat at the wide end. Cross section nearly rhombic. Form ovoid, with one flat end; slightly compressed front to back. The ventral face is strongly convex, with a central lengthwise line, the raphe. The dorsal face has 2 slightly convex halves that meet at approximately a right angle. There is a rim where the dorsal and ventral faces meet. On the ventral face there is a large flat area, the chalaza, near the small end. In the center of this area is the conspicuous caruncle. It is heart shaped, pointing toward the small end of the seed, ±0.6 mm wide, and yellowish with thin transparent yellow margins.

Surface strongly wrinkled and warty; dull. At high magnification a fine colliculate pattern can be seen over the entire surface. Brown-black; the high relief areas are brown, and the low areas and the raphe are black.

Length 2.5–2.7 mm; width 2.3–2.6 mm; thickness 2.1–2.3 mm.

Waste places, gardens, cultivated fields; good soils; S 2/3 GP.

125. *Euphorbia esula* L.   Leafy spurge

Seed: Outline ovate or rounded oblong with a tip at one small end. Cross section round. Form ovoid, rounded, with distinct dorsal and ventral faces. The dorsal face is convex with a small rounded tip. The ventral face of the seed has 2 flat or slightly convex halves separated by a lengthwise line, the raphe. At the small end of the ventral face is the depressed chalaza area. The elliptical hilum is in the depression. There is a very small caruncle at the end of the hilum near the tip of the seed. The raphe runs from the hilum to the center of the broad end of the seed.

Surface smooth; somewhat glossy. With magnification a fine reticulate pattern is visible. Pale gray-green with small yellow-brown flecks;

the raphe is dark. The seed appears to have a dark coat overlaid with a thin light-colored layer. The surface becomes darker with maturity.

Length 2.1–2.4 mm; width 1.7–1.8 mm; thickness 1.6–1.7 mm.

Pastures, cultivated fields, roadsides, waste places; open areas; N 2/3 GP.

126. *Euphorbia maculata* L.   Spotted spurge

Seed: Outline oblong, with one end tapered to a blunt tip. Cross section 4-angled; 2 sides meet at a right angle, and the opposite 2 sides form a broader angle. Form oblong, 4-angled, 4-sided. The faces are slightly concave. The right angle has a distinct convex curve and ends in a small tip. The 2 faces that form the right angle are larger than the other 2 faces. Between the 2 smaller faces is a narrow dark line, the raphe.

Surface with about 3 large but faint crosswise wrinkles. The surface appears to have a dull whitish outer layer that has a very fine alveolate texture. Red-brown with a grayish or whitish cast.

Length 0.8–0.9 mm; width 0.6 mm; thickness 0.5–0.6 mm.

Lawns, roadsides, waste places; disturbed soil; SE 1/4 GP.

127. *Euphorbia marginata* Pursh   Snow-on-the-mountain

Seed: Outline obovate. Cross section round. Form obovoid; a rim around the small end separates the ventral and dorsal faces. The rim is ±1.7 mm long. On the ventral side a distinct line, the raphe, extends from the rim to the large end. On the dorsal face there is a slight lengthwise central ridge. The hilum is on the small end of the ventral face, just below the rim.

Surface wrinkled; under high magnification it also appears finely roughened. There seems to be a thin glaucous layer over a dark surface. Dark grayish brown; less mature seeds are greenish or whitish.

Length 3.2–3.5 mm; thickness 2.6–2.8 mm.

Pastures, roadsides; calcareous soil, disturbed areas; throughout GP except N 1/4.

128. *Euphorbia nutans* Lag.   Eyebane

Seed: Outline ovate. Cross section 4-sided, almost square, but with 1 rounded corner. Form oblong, tapered; 3 of the angles form smooth ridges, whereas the fourth, more rounded angle has a narrow groove, the raphe. On this angle, near the small end of the seed, is the small indistinct hilum and a tiny caruncle.

Surface wrinkled crosswise; dull. The wrinkles are barely visible without magnification but are distinct at 4×. On each face there are 4 or 5 crosswise wrinkles as well as some in other directions. Under magnification the entire surface appears finely granular and has a slight sheen. Brownish gray; the angles are paler, and the raphe dark.

Length 1.3–1.4 mm; thickness 0.8–0.9 mm.

Waste areas, cultivated fields, roadsides; moist soil; SE 1/4 GP.

129. *Euphorbia prostrata* Ait. Creeping spurge, prostrate spurge

Seed: Outline bullet shaped, with 1 short straight end and 2 long gradually curved sides that meet in a blunt tip. Cross section square. Form an elongate, 4-sided, 4-angled shape with a flat base and bluntly pointed apex. It is broadest about 1/3 the distance from the base. The hilum is very small, located in the center of the base.

Surface with prominent crosswise wrinkles; dull. At low magnification 5 to 8 wrinkles can be distinguished on each face. At high magnification the surface also appears finely striate and dull. Light gray-brown; the wrinkles and angles are light gray, and the low areas are light brown.

Length 1.0–1.1 mm; thickness 0.5–0.6 mm.

Waste areas, lawns, roadsides; disturbed soil; S 1/2 GP except W 1/4.

130. *Tragia betonicifolia* Nutt. Nettleleaf noseburn

Seed: Outline round. Cross section round. Form nearly spherical, with 1 flat area. The hilum, in the flat area, is an irregular elliptical scar, ±0.8 mm long. The raphe, a narrow ridge ±1 mm long, extends from one end of the hilum. There is a barely detectable ridge around the circumference of the seed, from the raphe to the hilum.

Surface smooth; slight sheen. When seen at 10× magnification it appears very finely granular and shiny; at high magnification it appears colliculate. Mottled light and dark red-brown or light and dark gray-brown; the hilum is whitish, and the area around the hilum usually has a red or orange tinge.

Length 3.7–3.9 mm; width 3.4–3.6 mm; thickness 3.3 mm.

Pastures, waste places; open rocky hillsides; SE 1/8 GP.

131. *Tragia ramosa* Torr. Noseburn

Seed: Outline round. Cross section round. Form spherical. The hilum is an irregular elliptical scar, ±1 mm long. The raphe, a narrow ridge ±0.8 mm long, extends from one end of the hilum.

Surface smooth; dull. When seen at 10× magnification it appears finely granular; at high magnification it appears colliculate and resinous. Red-brown, orange, and buff in large mottled patches; the hilum is whitish, and the area around the hilum is orange.

Diameter 3.1–3.4 mm.

Pastures, waste places; open rocky hillsides; S 1/3 GP.

Sapindaceae Soapberry Family

132. *Cardiospermum halicacabum* L.

Common balloon vine

Seed: Outline round. Cross section round. Form spherical. The attachment area is a large cordate patch, which is dull and pale. It covers 1/3 or more of the surface. The hilum is an indistinct spot near the notch of the attachment area.

Surface (dark area) is smooth, glaucous, gray-black. The attachment area is dull, buff-colored. (The attachment area is a thin corky layer over the dark seed coat; the corky layer can be scraped off.)

Diameter 4.7–5.1 mm.

Waste places; moist soil; SE 1/8 GP.

Anacardiaceae Cashew Family

133. *Rhus glabra* L.

Smooth sumac

Seed: Outline ovate or short elliptical. Cross section ovate. Form ovoid or ellipsoid, slightly compressed. The hilum is indistinct, elliptical, ±1 mm long, located on one side margin, below the center. A faint seam-like marginal stripe extends from the hilum toward the farther end of the seed and 2/3 the length of the opposite side.

Surface nearly smooth; dull. At high magnification, both scurfy patches and shiny areas are visible. Dull light brown; the hilum is whitish.

Length 3.0–3.3 mm; width 2.3–2.5 mm; thickness 1.8–2.1 mm.

Roadside and pasture thickets; dry soil; E 1/2 GP.

Zygophyllaceae Caltrop Family

134. *Tribulus terrestris* L.

Puncture vine

Mericarp (an indehiscent segment of a schizocarp): Outline generally oblong with many short spines and 1 or 2 long spines. Cross section a wedge shape with 2 stout spines spreading at a wide angle from the corners of the broad end. Form sectorlike, with 2 flat faces that meet at an angle of 45° or less; it is flattened at the base and narrowed toward the apex. A ±3–5-mm-long spine flares from the midpoint of the outer margin of each flat face.

Surface conspicuously sculpted with many thin ridges and prominent tips; dull. Each tip has a long white bristle, but these are easily broken off. The entire surface has a sparse covering of fine short white hairs. Tan.

Length 4.7–5.7 mm; width 3.5–6.0 mm; thickness 1.8–2.5 mm. (Length measured along the inner angle of the sector.)

Roadsides; disturbed soil, often sandy or gravelly soil; S 2/3 GP.

Oxalidaceae
Wood Sorrel
Family

135. *Oxalis dillenii* Jacq. **Gray-green wood sorrel**

Seed: Outline elliptical. Cross section biconvex. Form ellipsoid, compressed, with 1 blunt end and 1 pointed end. The inconspicuous hilum is located at the pointed end of the seed.

Surface with sharply defined ridges that are clearly visible at low magnification; dull. There are about 9 crosswise ridges on each broad face; they are somewhat irregular and broken. There are also slight lengthwise wrinkles. The side margins sometimes include a narrow smooth groove. Dark orange-brown with off-white ridges. (Immature seeds have glossy red-brown ridges with dull pale intervals.)

Length 1.1–1.2 mm; width 0.8–0.9 mm; thickness ±0.4 mm.

Lawns; moist soil; throughout GP except SW 1/4.

136. *Oxalis stricta* L. **Yellow wood sorrel**

Seed: Outline elliptical with an acuminate tip at one end. Cross section an oblong with convex sides. Form ellipsoid with 1 pointed end; compressed. The hilum is at the pointed end.

Surface conspicuously ridged; dull. There are about 9 sharp crosswise ridges on each face; these ridges are bent or interrupted by 2 or 3 faint lengthwise wrinkles. The margins of each broad face are narrowly rimmed, and each long edge has a groove or smooth area between the rims. At 10× magnification the ridges are shiny and the interspaces dull. Dark orange-brown.

Length 1.3–1.4 mm; width 0.8–0.9 mm; thickness 0.4–0.5 mm.

Waste areas, lawns; moist soil; E 1/3 GP.

Geraniaceae
Geranium Family

137. *Geranium carolinianum* L. **Carolina cranesbill,**
wild geranium

Seed: Outline ovate or rounded oblong with a very small tip on one end. Cross section nearly round. Form oblong, rounded. The hilum is very narrow, ±0.7 mm long, and extends lengthwise from the center of one short side of the seed. Each end of the hilum projects as a slight tooth. From the hilum a narrow ridge runs to the far end of the seed.

Surface smooth; dull. With magnification a fine alveolate pattern is visible. Dark red-brown or gray-brown; the raised edges of the alveolate pattern are paler.

Length 2.2–2.3 mm; width 1.4–1.6 mm; thickness 1.4–1.5 mm.

Waste areas, lawns; good soils; SE 1/4.

**Apiaceae
Parsley Family**

138. *Cicuta maculata* L.   **Water hemlock**

Mericarp (fruit a schizocarp composed of 2 mericarps): Outline elliptical, with a flared apical extension. Cross section with 1 flat side and 1 strongly convex side. Form ellipsoid; ventral face flat and dorsal face strongly convex, with 5 conspicuous ribs. On the ventral face the attachment area (where the mericarps are attached to each other) is a central ovate patch that is slightly raised. This area is surrounded by a wide border of corky tissue. There is a narrow lengthwise line on the center of the ventral face. The stylopodium is a conspicuous extension from the apex.

Surface ribbed; dull. The ribs are smooth; the intervals appear granular or finely wrinkled. The ribs are yellow-brown to orange-brown and distinctly lighter than the brown to dark brown intervals. The raised area of the ventral face and the intervals between the ribs on the dorsal face are soft; they are oily when crushed. The raised area on the ventral face is brown with a buff border.

Length 2.7–3.3 mm; width (across the ventral face) 1.7–2.0 mm; thickness 1.4–1.8 mm.

Roadside ditches, stream and pond margins; wet soil; throughout GP except SW 1/4.

139. *Conium maculatum* L.   **Poison hemlock**

Mericarp (fruit a schizocarp composed of 2 mericarps that may be found singly or still joined): Outline ovate to long ovate, or plano-convex, with a conspicuous projection at the small end. Cross section with 1 straight side and 1 strongly convex side; the convex side has 5 distinct teeth. Form ovoid, flattened on the ventral side; the entire seed is slightly bent toward the flat face. There is a lengthwise central groove on the flat face. On the convex face there are 5 conspicuous lengthwise ribs. At the small end there is an irregular flaring tooth that is a remnant of the stylopodium.

The convex face is strongly ribbed. The ribs are smooth and may be straight or wavy. There are fine lengthwise wrinkles between the ribs. The surface is dull. Gray-brown; the ribs and stylopodium are paler and yellowish.

Length 2.5–3.0 mm; width (across the ventral face) 1.4–1.9 mm; thickness 1.1–1.5 mm.

Pastures, waste places, ditches; moist or shaded areas; central and E central GP.

140. *Daucus carota* L.   **Queen Anne's lace**

Mericarp (fruit a schizocarp composed of 2 mericarps, which may be found singly or still joined): Outline elliptical. Cross section like a disk with a missing sector, varying from about 3/5 to 3/4 of a circle. Form ellipsoid, curved, with a longitudinal sector removed. There is a deep lengthwise groove in the center of the concave (ventral) side. At one end this groove extends to form a small bifid projectlon, the stylopodium.

Surface bristly with closely spaced blunt curved spines. The spines have papillate bases and are minutely retrorsely barbed. Among the spines on the convex face of the seed are 3 narrow lengthwise rows of appressed white hairs. Some mericarps also have hairs along the ventral groove. Surface dull when seen without magnification. Under high magnification the spines have a glassy luster. Yellowish or grayish pale brown.

Length 3.2–4.2 mm; width 1.7–2.6 mm; thickness 1.3–2.0 mm. (Measurements include the brlstles.)

Roadsides, pastures; dry disturbed soil; E central GP.

Asclepiadaceae Milkweed Family

141. *Asclepias subverticillata* (A. Gray) Vail   **Poison milkweed**

Seed: Outline pear shaped with a truncate small end. Cross section a curved line with a distinct hump in the center. Form like a very thin, flattened pear. The whole seed may be rolled around the lengthwise axis. A large central area on each face is surrounded by a thin marginal wing. One face is distinctly convex. On the flatter face of the seed, the center of the seed coat appears to be adherent to the embryo, and there is a fine central lengthwise nerve from the truncate end toward the center. There is a readily shed coma of many white silky hairs, ±2.7 cm long, arising from the truncate end.

Surface smooth on the central convex area but slightly wrinkled on the margins and the flat face. There is a papery sheen. The overall color is light red-brown, with the convex area slightly darker than the margin. On the flat face there is a central narrow dark brown area, surrounded by a yellow-brown area and a medium brown wing.

Length 6.8–7.1 mm; width 3.7–4.2 mm; thickness (central area) ±0.7 mm; thickness (wing area) ±0.2 mm.

Rangeland, roadsides; sandy or rocky soil; SW 1/8 GP.

142. *Asclepias syriaca* L.   Common milkweed

Seed: Outline pear shaped. Cross section a narrow ellipse or dish shape. Form like a thin, flattened pear, with a truncate small end and a very thin marginal wing. The whole seed is somewhat concave-convex. On the dorsal face the location of the embryo is visible as a convex central area, also pear shaped. This area is surrounded by a thin wing that has an inner flat band, ±0.3 cm wide, and an outer, slightly wavy band, ±0.7 mm wide. On the ventral side there is a fine central line from the truncate end to near the wide end of the seed. There is a coma of many fine white silky hairs, ±3–3.5 cm long, spreading from the truncate end. The coma is readily shed, usually as a unit.

Surface papery; dull. On each face are many fine irregular wrinkles that are short but distinct. These are visible without magnification. At high magnification the surface of the outer bands of the wings appears finely colliculate. The overall color is medium orange-brown. The dorsal face is orange-brown with a slightly darker outer band. On the ventral face, the embryo area is light orange with a central dark brown elongate area; the inner band is dark orange-brown, and the outer band is the same color as on the dorsal side. The wrinkles are dark.

Length 7.0–8.5 mm; width 4.1–5.2 mm; thickness ±0.7–1.0 mm.

Roadsides, pastures, waste areas; open areas, often on clay soil; central and E central GP.

143. *Asclepias verticillata* L.   Whorled milkweed

Seed: Outline long-ovate or pear shaped, with a truncate small end. Cross section a very thin dish shape, somewhat thicker in the center. Form thin, ovate with a very thin, slightly wavy marginal wing. The wing is ±0.6–0.9 mm wide. The whole seed is somewhat bent or rolled around the lengthwise axis. The boundary between the central area and the wing is distinct on the convex face, with the center appearing swollen. On the concave face there is a fine central nerve extending from the truncate end 2/3 the length of the seed. There is a readily shed coma of many silky white hairs, ±2.5 cm long, arising from the truncate end.

Surface papery, slightly wrinkled; dull. The central area of each face is orange-brown. The wings are light orange-brown; often there is a band of dark brown on the inner part of the wings.

Length 5.1–6.2 mm; width (measured while rolled) 2.2–3.5 mm; thickness (of center area) ±0.4–0.5 mm.

Rangeland; open areas; throughout GP except SW 1/4.

144. *Cynanchum laeve* (Michx.) Pers.

Sand vine, climbing milkweed, honey vine

Seed: Outline ovate or pear shaped, with a truncate small end. Cross section a thin line with a slight thickening in the center. Form thin, ovoid, with a truncate small end and a conspicuous but readily deciduous coma. On one face the embryo is visible as a slightly elevated central area, ovate in shape. Around this elevated area is a broad wing that has a smooth inner band and a wavy outer band. On the other face of the seed the ovate central area is outlined but not elevated, and there is a narrow straight line that extends from the truncate end into the central area. The coma consists of many fine silky white hairs, 1.2–3.0 cm long.

Surface irregularly wrinkled; with a rough papery texture. Dark orange-brown. The outer band of the wing is a paler shade than the inner band and central area.

Length 5.4–9.0 mm; width 3.7–6.0 mm; thickness ±0.2 mm.

Fence rows, cultivated fields, roadsides; good soils; E central GP.

Solanaceae Potato or Nightshade Family

145. *Datura stramonium* L.

Jimson weed

Seed: Outline like a circle with a large notch or like most of a circle with 1 straight edge. Cross section rounded triangular. Form like a wedge, narrowed toward one end; there is a notch in the inner angle. The hilum is located in the notch, near the narrow end of the seed. The hilum is triangular, ±1.2 mm long and 0.7 mm wide, pointing to the narrow end of the seed.

Surface uneven, slightly ridged or wrinkled; dull. At 10× magnification a fine meshlike pattern is visible in addition to the larger wrinkles, and there is a slight sheen. Black; the area near the hilum is paler, and the hilum is orange. Immature seeds pale.

Length 3.4–3.8 mm; width 2.8–3.2 mm; thickness 1.4–1.7 mm.

Cultivated fields, barnyards; rich soil; E central and S 1/4 GP.

146. *Hyoscyamus niger* L.   Henbane

Seed: Outline variable, usually somewhat obovate. Cross section narrowly ovate. Form ovoid, compressed, asymmetrical, with one long edge thinner than the other; the broad faces flat to concave. The hilum is small, round, located at the small end of the seed.

Surface appears tuberculate at low magnification. At 10× magnification it is distinctly reticulate. The reticulum is of irregular height, and it extends outward into many blunt points or tubercles. Surface dull when seen without magnification; at 10× the low areas are very shiny. Light gray-brown, the ridges darker than the depressions. At high magnification the depressions appear pale with a fine dark branching pattern.

Length 1.4–1.6 mm; width 1.2–1.4 mm; thickness 0.7–0.8 mm.

Roadsides, pastures; open areas; N 1/2 GP.

147. *Physalis heterophylla* Nees   Clammy ground cherry

Seed: Outline variable, generally ovate. Cross section narrowly elliptical. Form like a thin chip of irregular thickness. The edges are rounded, and the whole seed may be slightly bent. The hilum is a narrow ellipse, ±0.7 mm long, on the margin of one long edge of the seed.

The surface is finely roughened; glossy. When seen with magnification it appears alveolate or pitted. Orange-yellow.

Length 1.8–2.4 mm; width 1.6–1.8 mm; thickness 0.2–0.5 mm.

Roadsides, gardens, waste areas; good soils;. throughout GP except SW 1/4.

148. *Physalis longifolia* Nutt.   Common ground cherry

Seed: Outline variable, but rounded and generally ovate. Cross section elliptical. Form compressed, ovoid, thickness irregular, with a narrow marginal rim. The hilum is small and narrow, located in a small notch in the rim.

Surface smooth; dull. At 10× magnification the surface appears to be covered with shiny granules. At high magnification these appear to be closely spaced tubercles. Dull orange-yellow.

Length (1.5) 1.9–2.2 mm; width (1.2) 1.5–1.8 mm; thickness 0.6–0.8 mm.

Roadsides, gardens, waste areas; moist soil; throughout GP except NE 1/8.

149. *Solanum carolinense* L.   Carolina horse-nettle

Seed: Outline somewhat ovate but variable. Cross section narrowly elliptical. Form compressed, ovoid, thickness irregular; the whole seed may be bent or curved. There is a shallow notch, ±1 mm long, along one edge near the smaller end of the seed. The hilum is a deep elliptical pit in this notch.

Surface smooth; glossy. When seen with magnification an alveolate pattern is visible. Yellow to orange-brown.

Length 1.9–2.8 mm; width 1.4–2.1 mm; thickness 0.6–0.8 mm. (The size varies considerably, with many apparently fully developed seeds smaller than the most common size, which is 2.7–2.8 mm in length.)

Pastures, roadsides; often on sandy soil; SE 1/4 GP.

150. *Solanum eleagnifolium* Cav.   Silver-leaf nightshade

Seed: Outline variable, but usually ovate, 3-sided, or oblong. Cross section narrowly elliptical. Form a flat thin chip with smooth rounded edges. Usually both faces are slightly convex. There is a shallow notch in the margin near one small end of the seed; the hilum is a deep ovate cavity in this notch.

Surface smooth; semiglossy. At high magnification it appears very finely textured. Medium or dark yellow-brown.

Length 2.6–3.5 mm; width 1.9–2.5 mm; thickness 0.6–0.8 mm.
Pastures, cultivated fields, waste areas; good soils; S 1/3 GP.

151. *Solanum ptycanthum* Dun. ex DC.   Black nightshade

Seed: Outline asymmetrically obovate, with a blunt tip at the small end. Cross section an irregular ellipse. Form a rounded chip of irregular thickness. The inconspicuous hilum is located just below the tip on the straighter edge of the seed.

Surface smooth; dull. When seen at 7× magnification, it appears finely pitted and shiny. At high magnification there appears to be a translucent shiny mesh laid over an opaque seed. Yellow-brown to orange-brown; the rim may appear darker.

Length 1.7–2.1 mm; width 1.4–1.6 mm; thickness 0.6–0.7 mm.
Gardens, cultivated fields, especially soybeans and field beans, waste places; fertile soil; throughout GP except W 1/4.

152. *Solanum rostratum* Dun. **Buffalo bur**

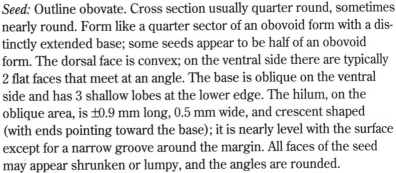

Seed: Outline irregularly ovate, with one side nearly straight and slightly notched. Cross section a narrow oblong. Form flattened ovoid. The hilum is in the notch.

Surface finely reticulate at 7× magnification; dull. When seen at 10× magnification, the raised edges of the meshwork appear shiny or light colored, whereas the pits are very dark. Dark red-brown to black.

Length 2.4–2.7 mm; width 2.0–2.1 mm; thickness 0.9–1.0 mm.

Pastures, crop fields, feed lots, roadsides; good soils, often in dry areas; throughout GP.

153. *Solanum triflorum* Nutt. **Cut-leaved nightshade**

Seed: Outline variable, but usually elliptical, with a short straight edge near one end of the ellipse. Cross section narrowly elliptical. Form ellipsoid, compressed, thickness irregular. The 2 faces are slightly convex, and the margins are thin. The hilum, located on the straight part of the margin, is very narrow.

Surface finely roughened; dull. At high magnification small irregular tubercles are visible. Yellow-brown to brown.

Length 2.0–2.2 mm; width 1.2–1.5 mm; thickness 0.4–0.5 mm.

Pastures, cultivated fields, especially bean fields; rocky soil; throughout GP.

Convolvulaceae Morning Glory Family

154. *Calystegia sepium* (L.) R. Br. subsp. *angulata* Brummitt **Hedge bindweed**

Seed: Outline obovate. Cross section usually quarter round, sometimes nearly round. Form like a quarter sector of an obovoid form with a distinctly extended base; some seeds appear to be half of an obovoid form. The dorsal face is convex; on the ventral side there are typically 2 flat faces that meet at an angle. The base is oblique on the ventral side and has 3 shallow lobes at the lower edge. The hilum, on the oblique area, is ±0.9 mm long, 0.5 mm wide, and crescent shaped (with ends pointing toward the base); it is nearly level with the surface except for a narrow groove around the margin. All faces of the seed may appear shrunken or lumpy, and the angles are rounded.

Surface finely roughened or scurfy at low magnification; dull. At high magnification there are scattered warty projections on a finely textured background. Dark brown to black; the hilum is orange.

Length 4.6–5.1 mm; width 3.2–3.7 mm; thickness 3.0–3.6 mm.
Roadsides, fence rows; deep or moist soil; throughout GP except SW 1/8.

155. *Convolvulus arvensis* L.   Field bindweed

Seed: Outline obovate with a truncate small end. Cross section quarter round. Form like a quarter sector of an obovoid form; typically there are 2 flat faces at right angles on the ventral side and a strongly convex dorsal face. Some seeds have an angle greater than 90° between the flat faces. The faces sometimes appear shrunken. At the small end there is a small flat area, oblique or at a right angle to the main angle of the seed. The hilum, placed crosswise in this flat area, is oblong, ±0.5 mm long.

Surface uniformly covered with small rough irregular tubercles; very dull. Dark gray-brown; less mature seeds are reddish.

Length 3.3–3.7 mm; thickness 0.7–3.0 mm.

Cultivated fields, especially grain fields and alfalfa, gardens, old fields, waste places, roadsides; moist or dry soil in open areas; throughout GP.

156. *Ipomoea hederacea* Jacq.   Ivyleaf morning-glory

Seed: Outline semicircular. Cross section a wedge, varying from about 1/6 to 1/4 of a circle, with a depression in the center of the curved edge. Form a sector, with 2 flat ventral faces that meet at an angle of 90° or less, and a convex dorsal face that has a central furrow. There is a slight rim around the edge of the convex face. The entire seed appears somewhat wrinkled and irregular. The hilum is located in a shallow notch that occupies the lower third of the angle formed by the flat faces.

Surface finely granular or scurfy; dull. At 7× magnification a covering of short fuzzy brown hairs is visible. The hairs are more abundant near the angles. Dark brown-black.

Length 4.8–5.6 mm; width 3.4–4.2 mm; thickness 3.3–4.5 mm.

Cultivated fields, roadsides; moist or dry soil; SE 1/4 GP.

157. *Ipomoea lacunosa* L.   White morning-glory

Seed: Outline rounded triangular. Cross section a wedge, with an angle slightly less than 90°. Form a quarter sector, with 2 nearly flat faces on the ventral side and a broad, curved dorsal face that is distinctly bowed out in the middle and flat toward the base and apex. The faces usually appear lumpy or shrunken. A shallow notch in the angle extends about 1/3 the distance from the base. The hilum, in the notch, is circular, ±0.9 mm in diameter, and level with the surface except for a groove around its circumference.

Surface smooth; soft sheen. Black, tinged with red-brown.

Length 4.2–4.7 mm; width (across the dorsal face) 3.7–4.8 mm; thickness 3.5–3.6 mm.

Cultivated fields; moist soil; S 1/2 except W 1/4.

158. *Ipomoea purpurea* (L.) Roth   Common morning-glory

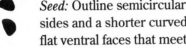

Seed: Outline semicircular. Cross section a wedge with 2 long straight sides and a shorter curved or indented side. Form a sector, having 2 flat ventral faces that meet in an angle and a rounded dorsal face that sometimes has a central furrow. The entire seed is somewhat wrinkled and irregular in form. There is a flattened area at one end of the angle; the hilum, ±0.7 mm in diameter, is in this area.

Surface densely covered with tiny brownish transparent hairs and transparent granular processes, which are visible at 10× magnification; dull. At lower magnification the surface may appear granular. Black to brown-black.

Length 4.6–5.1 mm; width (of dorsal face) 2.8–4.1 mm; thickness 3.4–4.0 mm.

Cultivated fields, gardens; disturbed soil; E 1/4 and central GP.

Cuscutaceae
Dodder Family

159. *Cuscuta indecora* Choisy   Large alfalfa dodder

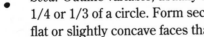

Seed: Outline variable, usually asymmetrically round. Cross section is 1/4 or 1/3 of a circle. Form sectorlike, with a convex dorsal face and 2 flat or slightly concave faces that meet in an angle on the ventral side. The angle is most commonly ±120° but may be smaller or larger; a few seeds are hemispheric. The edge where the convex face joins the flat faces is very rounded. The ventral side is obliquely truncate at the base. The hilum, in the truncate area, is a circular patch with a small whitish scar in the center.

Surface finely granular and scurfy; dull. At high magnification a fine reticulate pattern can be seen. Dull yellow to orange-brown.

Length 1.5–1.8 mm; width 1.2–1.5 mm; thickness 0.9–1.2 mm.

Parasitic on many broad-leaved plants, including crops; throughout GP.

160. *Cuscuta pentagona* Engelm.   Field dodder

Seed: Outline ovate, often with an obliquely truncate base. Cross section ovate or like a broad wedge. Form sectorlike and irregular or ovoid and compressed. The dorsal face is convex, and the ventral side may be nearly flat or composed of 2 flat faces that form a large angle. The hilum, ±0.3 mm in diameter, is in the truncate or notched area at the base of the ventral side.

Surface smooth; dull. At high magnification it appears finely colliculate. Orange-brown.

Length 1.1–1.2 mm; width (across the dorsal face) 0.9–1.1 mm; thickness 0.6–1.0 mm.

Parasitic on legume crops and other broad-leaved plants; throughout GP except SW 1/8.

Hydrophyllaceae Waterleaf Family

161. *Ellisia nyctelea* L.   Waterpod

Seed: Outline round. Cross section round. Form globose but with 1 somewhat flattened area. The hilum is very small and inconspicuous.

Surface smooth; dull. At 7× magnification a fine alveolate pattern is visible; the ridges are dull, and the interspaces appear glittery or shiny. Black.

Diameter 2.2–2.8 mm.

Cultivated fields, gardens; moist soil; throughout GP except SW 1/4.

Boraginaceae Borage Family

162. *Cynoglossum officinale* L.   Hound's tongue

Nutlet: Outline obovate. Cross section elliptical. Form obovoid, compressed. Dorsal face flat; ventral face convex, with a large obovate scar at the small end. The small round hollow attachment point is located off center on the scar. The scar is ±3.5 mm long, 2.0 mm wide; its small end extends into a free strip of tissue, ±2.0 mm long.

Surface, except scar, closely covered by stout spines that are tipped with small downwardly directed barbs. The surface between the spines is smooth. Dull. The scar area is papery, rough, dull. Light gray-brown; the scar is light brown. The apical barbs on the spines are shiny and transparent.

Length 6.1–7.3 mm; width 4.9–6.5 mm; thicknesss 2.8–3.5 mm. (Measurements include the spines.)

Pastures, roadsides; disturbed soil; N 1/2 and E central GP.

163. *Lithospermum arvense* L.   Corn gromwell

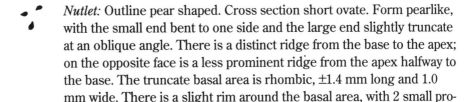

Nutlet: Outline pear shaped. Cross section short ovate. Form pearlike, with the small end bent to one side and the large end slightly truncate at an oblique angle. There is a distinct ridge from the base to the apex; on the opposite face is a less prominent ridge from the apex halfway to the base. The truncate basal area is rhombic, ±1.4 mm long and 1.0 mm wide. There is a slight rim around the basal area, with 2 small projections aligned with the ridges.

Surface covered with small tubercles and wrinkles; shiny. Most of the wrinkles are oriented lengthwise. The base is dull and finely textured. Light gray-brown overall; the ridges, wrinkles, and tubercles are light colored, and the low relief areas and the base are darker.

Length 2.2–2.8 mm; thickness 1.3–1.8 mm.

Winter wheat fields, waste areas; sandy soil; SE 1/4 GP.

Verbenaceae
Vervain Family

164. *Verbena bracteata* Lag. & Rodr.   Prostrate vervain

Nutlet: Outline a rounded rectangle. Cross section like a quarter sector with a bulging curved side. Form oblong with 2 flat faces and 1 larger, strongly convex face; the edges of the convex face are distinctly rimmed. The attachment scar is a quarter-round area at the base.

Surface of the convex face is strongly reticulate and has 3 or 4 low lengthwise ridges; there are also a few irregular crosswise ridges, mostly near the basal end. When seen at 10× magnification, the flat faces appear to have a dense covering of white or reddish granules. At high magnification these are visible as distinct barblike processes. Surface dull. Brown, with a whitish attachment scar.

Length 1.8–2.2 mm; thickness (across the convex face) 0.6–0.7 mm.

Roadsides, pastures; dry, compacted soil; throughout GP.

165. *Verbena hastata* L.   Blue vervain

Nutlet: Outline elliptical. Cross section quarter round. Form an elongate quarter sector. The dorsal face is convex and has a narrow, flaring marginal wing. On the ventral side are 2 flat faces that meet at a right angle, with a slight ridge along the angle. The attachment scar is a conspicuous whitish spot near the base on the ventral side.

Surface nearly smooth; somewhat shiny. On the convex face there are 3 faint lengthwise ridges as well as several faint crosswise wrinkles near the apex. The flat faces may appear scurfy, with scattered white flecks. At high magnification these flecks appear to be short, thick appressed hairs. Orange-brown with a translucent marginal wing.

Length 1.7–1.9 mm; width 0.7–0.8 mm; thickness 0.4–0.5 mm.

Pastures; moist soil; throughout GP except NW 1/8 and SW 1/8.

166. *Verbena stricta* Vent.   Hoary vervain

Nutlet: Outline a rounded oblong. Cross section quarter round. Form an elongate sector, with a convex dorsal face; on the ventral side are 2 flat faces that meet at a right angle. There is a thin rim all around the convex face. At the base of the angle on the ventral side is the hollow attachment area.

Surface of the convex face has about 5 narrow lengthwise ribs and a reticulate pattern of wrinkles. The wrinkles are more prominent toward the apical end. The spaces between the wrinkles are finely textured; semiglossy. The flat faces appear sparsely to densely scurfy with white flecks. At high magnification these flecks are seen to be very short appressed hairs that are opaque. Dark red-brown.

Length 2.7–3.1 mm; width 0.8–1.0 mm; thickness 0.7 mm.

Rangeland, pastures, roadsides; dry areas; throughout GP except NW 1/8 and SW 1/8.

167. *Verbena urticifolia* L.   Nettle-leaved vervain

Nutlet: Outline a rounded oblong, somewhat narrower at the base than the apex. Cross section quarter round. Form an elongate sector with a convex dorsal face; on the ventral side are 2 flat faces that meet at a right angle. The margin of the convex face has a narrow rim. The attachment point is a quarter round area at the base.

Surface smooth; dull. With magnification, the convex face appears shiny and has about 3 faint lengthwise ribs and a few irregular crosswise wrinkles, whereas the flat faces appear slightly granular or scurfy. Dull orange-brown.

Length 3.3–3.6 mm; thickness (across the convex face) 1.5–1.6 mm.

Pastures; open areas; E 1/2 GP.

Lamiaceae
Mint Family

168. *Glecoma hederacea* L.   Ground ivy

Nutlet: Outline obovate. Cross section a quarter circle. Form an elongate sector. The dorsal face is convex; the ventral side has 2 flat faces that meet in a lengthwise angle. The attachment area, ±0.2 mm in diameter, is a depression with a thin collar, near the base on the ventral side.

Surface smooth; slight sheen. Dull brown, with a faint dark midvein on the dorsal side and a dark area around the attachment scar.

Length 1.6–1.7 mm; width 1.0–1.1 mm, thickness 0.6–0.7 mm.

Lawns; moist soil; E 1/2 GP.

169. *Hedeoma hispidum* Pursh   Rough false pennyroyal

Nutlet: Outline elliptical. Cross section rounded triangular. Form ellipsoid, somewhat sectorlike. Dorsal side convex; ventral side with 2 faces that meet in a rounded angle of ±90°. The attachment area is a V-shaped depression at the basal end of the ventral side.

Surface alveolate; there is a slight sheen at 10× magnification. Brown; the attachment area is whitish and scurfy.

Length 1.0–1.1 mm; width across dorsal face 0.6 mm; thickness 0.5 mm.

Roadsides, waste areas; dry open soil; throughout GP except SW 1/8.

170. *Lamium amplexicaule* L.   Henbit

Nutlet: Outline long-obovate. Cross section a quarter circle. Form obovoid, sectorlike, with a convex dorsal face; on the ventral side are 2 flat faces that meet in a right angle. There is a narrow line along this angle. The apical end of the seed is obliquely truncate. At the basal end there are ridges along the angles between the faces.

Surface smooth; dull. At high magnification a fine alveolate pattern is visible. Dark gray-brown but with many distinct white mottles that are more abundant toward the apex; the ridges are yellow.

Length 1.6–2.0 mm; width 0.9–1.0 mm; thickness ±0.7 mm.

Winter wheat fields, fallow fields, waste areas, gardens; moist soil; SE 1/4 and central GP.

171. *Salvia reflexa* Hornem.   Lance-leaved sage

Nutlet: Outline elliptical, with one end more rounded than the other. Cross section a quarter circle. Form an elongate sector. The dorsal side is convex; on the ventral side are 2 flat faces that meet in a rounded right angle. The basal end of the seed is slightly pointed. The attachment scar is a slightly elevated elongate patch near the base on the ventral side.

Surface smooth; dull. When seen at high magnification it is semi-glossy with scattered orange resinous dots. Tan with brown mottling; the attachment area whitish.

Length 2.2–2.3 mm; width 1.5–1.6 mm; thickness 1.0–1.1 mm.
Roadsides, pastures; rocky or sandy soil; throughout GP.

**Plantagniaceae
Plantain Family**

172. *Plantago lanceolata* L. Buckhorn plantain

Seed: Outline elliptical to long ovate. Cross section C-shaped. Form ellipsoid, compressed, slightly concave-convex, with the margins rolled
toward the concave side. The hilum is an elliptical scar placed lengthwise in the center of the concave face. From the ends of the hilum a
low ridge runs to each end of the seed.

Surface smooth when seen without magnification; at 10× it appears
finely roughened. The concave surface is dull, but the convex face and
the inrolled margins are shiny. Translucent, orange-brown; on the convex face there is a broad central lengthwise stripe that is paler and
opaque, and the hilum is dark or black.

Length 2.0–2.3 mm; width 1.0–1.2 mm; thickness 0.5–0.7 mm.
(Thickness measured from rim to high point of convexity.)

*Clover and alfalfa fields, lawns, pastures, waste places; good soils; E
central and central GP.*

173. *Plantago major* L. Common plantain

Seed: Outline variable, elongate, usually angular, often triangular or
rhombic. Cross section usually 3- or 4-sided. Form elongate, generally
ellipsoid but angular; somewhat compressed with one broad face often
nearly flat and the other biconvex to pyramidal. The hilum is indistinct, located at the tip of the pyramid or convexity.

Surface rough, uneven; dull. Brown-black to black.

Length 0.9–1.2 mm; width 0.5–0.7 mm; thickness 0.3–0.4 mm.

*Lawns, pastures, waste places; open areas; throughout GP except
S 1/4.*

174. *Plantago patagonica* Jacq. Patagonian plantain, woolly plantain

Seed: Outline long ovate. Cross section C-shaped. Form long-ovoid,
with a convex dorsal face and a concave ventral face. There is a faint
groove across the center of the convex face. On the concave face the
hilum is visible as 2 small depressions in the center.

Surface finely roughened; dull. At high magnification a colliculate
pattern is visible. Orange-brown with a translucent margin; on the convex face there is a paler central area, and on the concave face there is

an orange central area surrounded by an ovate white ring and an orange border.

Length 1.9–2.1 mm; width 1.0–1.1 mm; thickness 0.6–0.7 mm.
Pastures, roadsides; open dry areas; throughout GP.

175. *Plantago rugelii* Dcne.   Rugel's plantain

Seed: Outline variable; may be long triangular with either rounded or angular corners, an ellipse with one truncate end, or variously long and angular. Cross section a thin wedge or ellipse. Form chiplike, irregular, with 1 strongly convex face and 1 convex or slightly concave face. These faces meet in a narrow rim around most of the periphery, but there is a narrow marginal face on one short edge. The hilum is an elliptical scar in the center of the less convex face of the seed.

Surface very finely wrinkled; dull. At high magnification there appears to be a translucent, somewhat shiny surface layer. Very dark red-brown to nearly black; hilum light brown.

Length 1.5–2.0 mm; width 0.8–0.9 mm; thickness 0.4–0.5 mm.
Lawns, waste areas; shaded dry soil or heavily compacted soil; E 1/2 GP.

176. *Plantago virginica* L.   Pale-seeded plantain

Seed: Outline long ovate. Cross section a thin concave-convex lens shape. Form ovoid, compressed, thin, with margin extended toward one side to form 1 concave face and 1 convex face.

Surface smooth; the convex face is shiny, the concave face dull. Under magnification the surface appears very finely textured. Light orange-brown, translucent near the the margins.

Length 1.6–1.8 mm; width 0.8–1.0 mm; thickness 0.5–0.6 mm.
Pastures, lawns, waste places; open areas; SE 1/4 GP.

Scrophulariaceae
Figwort Family

177. *Verbascum blattaria* L.   Moth mullein

Seed: Outline variable, roughly rectangular or trapezoidal. Cross section irregular, rounded. Form cylindrical, or truncate conical, with a flat base and more rounded apex. The hilum is at the center of the base.

Surface rough; dull. There are 7 or 8 lengthwise rounded ridges and about 6 crosswise ridges that intersect to form deep crosswise pits. At 25× magnification a finer pattern of small, mostly lengthwise wrinkles is also visible. Fine striations, visible at 15×, radiate from the hilum and extend lengthwise over the whole surface. Dark gray-brown; the rim of the base is black, and there is a small black spot at the center of the apex and of the base.

Length 0.7–1.0 mm; diameter 0.4–0.7 mm.

Old fields, waste places; good soils; SE 1/4 GP.

178. *Verbascum thapsus* L.   Common mullein

Seed: Outline variable, oblong or elongate trapezoidal. Cross section round or angled. Form variable, somewhat like a truncate cone with an oblique base and a rounded apex. Hilum small, at the center of the base.

Surface conpicuously knobby; dull. There are 8 to 10 wavy length-wise ridges that are intersected by several irregular crosswise ridges, forming oblong projections separated by deep grooves. There is a very fine reticulate pattern overall. Dark gray-brown.

Length 0.7–0.9 mm; diameter 0.4–0.5 mm.

Pastures, old fields, roadsides, waste; disturbed or rocky soil; through-out GP except SW 1/8.

179. *Veronica agrestis* L.   Field speedwell

Seed: Outline ovate. Cross section dish shaped. Form thin, ovate, strongly concave-convex. The margins are sometimes rolled toward the concave face. The convex face has a faint central lengthwise ridge. In the center of the concave face there may be remnants of the funiculus.

Surface of the concave face smooth; the convex face is covered with crosswise wrinkles, clearly visible at 7× magnification. Surface dull. Pale yellow.

Length 1.3–1.6 mm; width 1.0–1.3 mm; thickness ±0.1 mm.

Lawns, waste places; open areas; SE 1/4 and central GP.

180. *Veronica arvensis* L.   Corn speedwell

Seed: Outline ovate with a notch at the small end. Cross section a thin dish or V-shape, with a small knob at the center of the concave side. Form like an ovate chip, somewhat concave-convex. Hilum a raised el-lipse, ±0.4 mm long, in the center of the concave face.

Surface covered with fine wrinkles visible at 7× magnification. Dull, but when seen with 10× magnification there is a waxy sheen. Pale dull orange, somewhat translucent; the hilum is darker.

Representative Seeds of Some Groups of Weedy Plants

Bean family
Fabaceae
4.8x

Buckwheat family
Polygonaceae
4.8x

Spurge family
Euphorbiaceae
4.8x

Thistles
Asteraceae
4.8x

Mustard family
Brassicaceae
9.6x

Morning Glory family
Convolvulaceae
4.8x

10   20   30

Seeds of Weedy Plants

1. *Ranunculus abortivus*
Early wood buttercup
9.6x

2. *Argemone polyanthemos*
Prickly poppy
4.8x

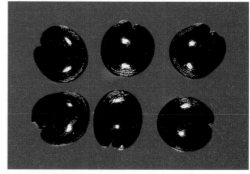

3. *Argemone squarrosa*
Hedgehog prickly poppy
4.8x

4. *Corydalis aurea*
Golden corydalis
9.6x

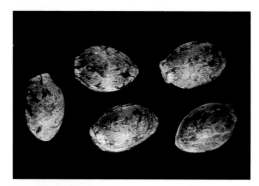

5. *Cannabis sativa*
Marijuana
4.8x

6. *Urtica dioica*
Stinging nettle
9.6x

10   20   30

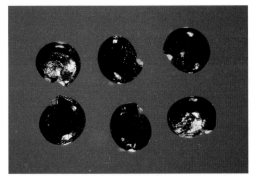

7. _Phytolacca americana_
Pokeweed
4.8x

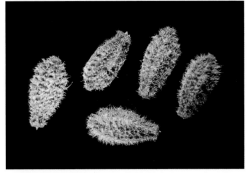

8. _Mirabilis nyctaginea_
Wild four-o'clock
4.8x

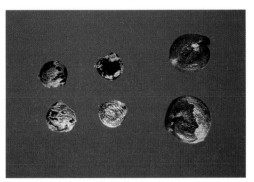

9. _Atriplex subspicata_
Spearscale
4.8x

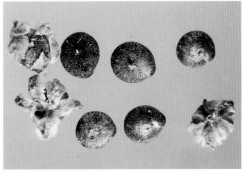

10. _Chenopodium album_
Lamb's quarters
9.6x

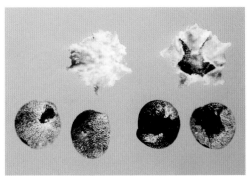

11. _Chenopodium berlandieri_
Pitseed goosefoot
9.6x

12. _Chenopodium gigantospermum_
Maple-leaved goosefoot
9.6x

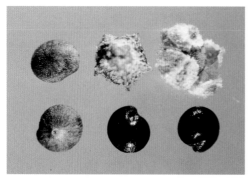

13. *Chenopodium pratericola*
Dryland goosefoot
9.6x

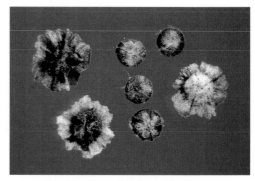

14. *Cycloloma atriplicifolium*
Tumble ringwing
4.8x

15. *Kochia scoparia*
Fireweed
9.6x

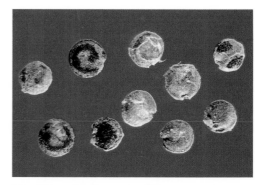

16. *Monolepis nuttalliana*
Poverty weed
9.6x

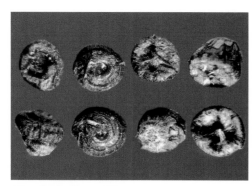

17. *Salsola collina*
Tumbleweed
9.6x

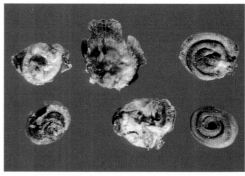

18. *Salsola iberica*
Russian-thistle
9.6x

10   20   30

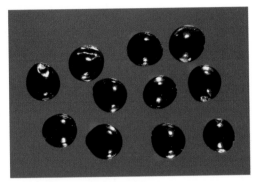

19. *Amaranthus albus*
Tumble pigweed
9.6x

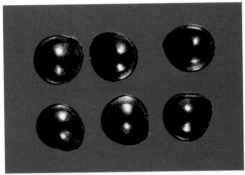

20. *Amaranthus graecizans*
Prostrate pigweed
9.6x

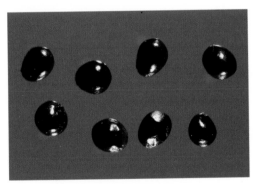

21. *Amaranthus hybridus*
Slender pigweed
9.6x

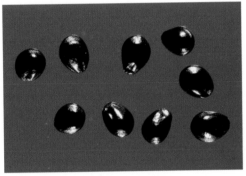

22. *Amaranthus palmeri*
Palmer's pigweed
9.6x

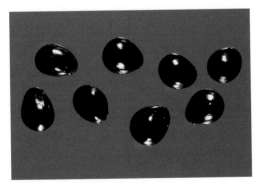

23. *Amaranthus retroflexus*
Rough pigweed
9.6x

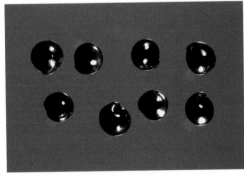

24. *Amaranthus rudis*
Water-hemp
9.6x

25. *Amaranthus spinosus*
Spiny pigweed
9.6x

26. *Froelichia floridana*
Field snake-cotton
9.6x

27. *Portulaca oleracea*
Purslane
9.6x

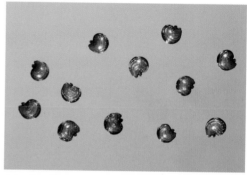

28. *Mollugo verticillata*
Carpetweed
9.6x

29. *Cerastium vulgatum*
Mouse-ear chickweed
9.6x

30. *Holosteum umbellatum*
Jagged chickweed
9.6x

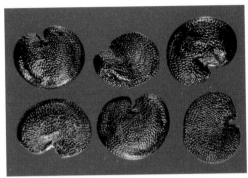

31. *Saponaria officinalis*
Soapwort
9.6x

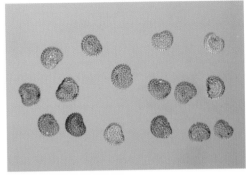

32. *Silene antirrhina*
Sleepy catchfly
9.6x

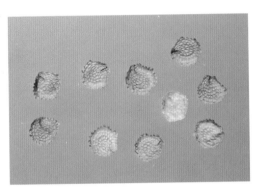

33. *Stellaria media*
Common chickweed
9.6x

34. *Vaccaria pyramidata*
Cow-cockle
9.6x

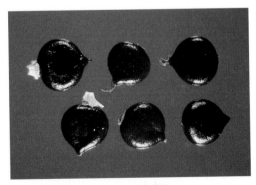

35. *Polygonum amphibium*
Swamp smartweed
4.8x

36. *Polygonum arenastrum*
Knotweed
4.8x

10   20   30

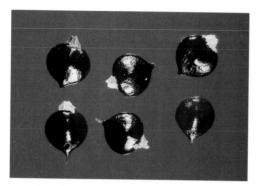

37. *Polygonum bicorne*
Pink smartweed
4.8x

38. *Polygonum convolvulus*
Wild buckwheat
4.8x

39. *Polygonum lapathifolium*
Pale smartweed
4.8x

40. *Polygonum pensylvanicum*
Pennsylvania smartweed
4.8x

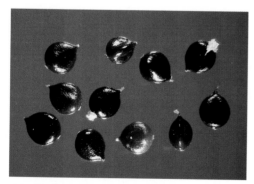

41. *Polygonum persicaria*
Lady's thumb
4.8x

42. *Polygonum ramosissimum*
Bush knotweed
4.8x

43. *Polygonum scandens*
Climbing false buckwheat
4.8x

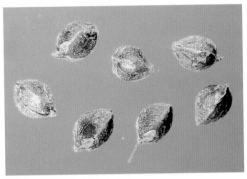

44. *Rumex acetosella*
Sheep sorrel
9.6x

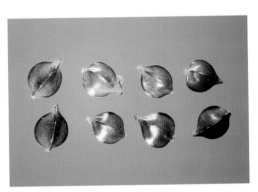

45. *Rumex altissimus*
Pale dock
4.8x

46. *Rumex crispus*
Curly dock
4.8x

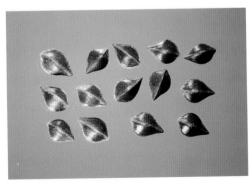

47. *Rumex mexicanus*
Willow-leaved dock
4.8x

48. *Rumex patientia*
Patience dock
4.8x

10   20   30

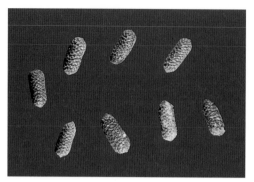

49. *Hypericum perforatum*
Common St. John's-wort
9.6x

50. *Abutilon theophrasti*
Velvet-leaf
4.8x

51. *Hibiscus trionum*
Flower-of-an-hour
4.8x

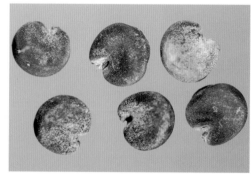

52. *Malva neglecta*
Common mallow
9.6x

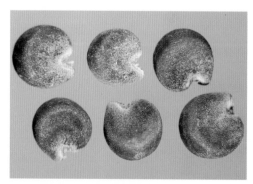

53. *Malva rotundifolia*
Round-leaf mallow
9.6x

54. *Malvastrum hispidum*
Yellow false-mallow
4.8x

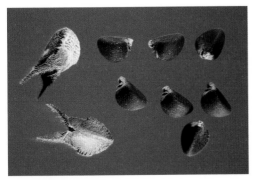

55. Sida spinosa
Prickly sida
4.8x

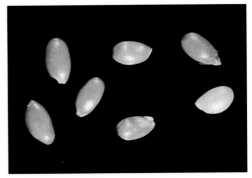

56. Viola rafinesquii
Johnny-jump-up
9.6x

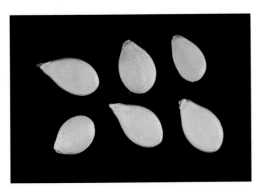

57. Cucurbita foetidissima
Buffalo-gourd
2x

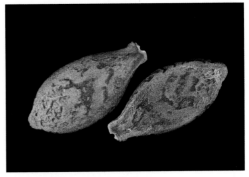

58. Echinocystis lobata
Wild cucumber
2x

59. Sicyos angulatus
Bur cucumber
2x

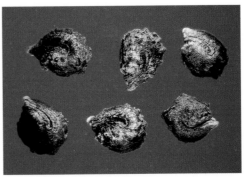

60. Cleome serrulata
Rocky Mountain bee plant
4.8x

10   20   30

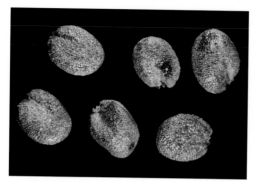

61. *Barbarea vulgaris*
Winter cress
9.6x

62. *Berteroa incana*
Hoary false alyssum
9.6x

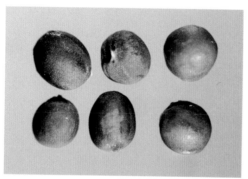

63. *Brassica juncea*
Indian mustard
9.6x

64. *Brassica kaber*
Charlock
9.6x

65. *Brassica nigra*
Black mustard
9.6x

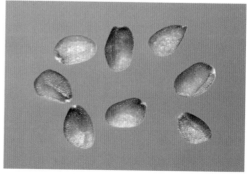

66. *Camelina microcarpa*
Small-seeded false flax
9.6x

67. *Camelina sativa*
Gold-of-pleasure
9.6x

68. *Capsella bursa-pastoris*
Shepherd's purse
9.6x

69. *Cardaria draba*
Hoary cress
9.6x

70. *Chorispora tenella*
Blue mustard
9.6x

71. *Conringia orientalis*
Hare's-ear mustard
4.8x

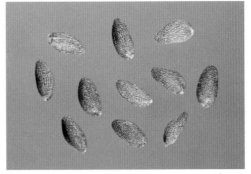

72. *Descurainia pinnata*
Tansy mustard
9.6x

|||||||||||||||||||||||||||||||||
10   20   30

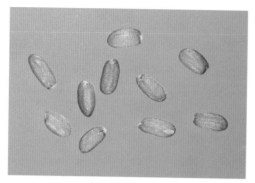

73. *Descurainia sophia*
Flixweed
9.6x

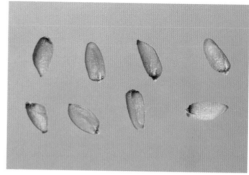

74. *Erysimum cheiranthoides*
Wormseed wallflower
9.6x

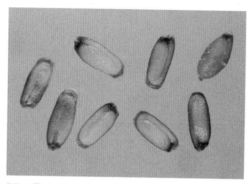

75. *Erysimum repandum*
Bushy wallflower
9.6x

76. *Lepidium campestre*
Field peppergrass
9.6x

77. *Lepidium densiflorum*
Peppergrass
9.6x

78. *Lepidium perfoliatum*
Clasping peppergrass
9.6x

10   20   30

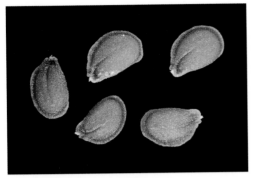

79. *Lepidium virginicum*
Virginia peppergrass
9.6x

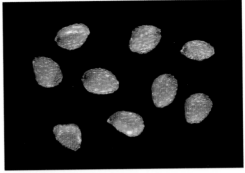

80. *Nasturtium officinale*
Watercress
9.6x

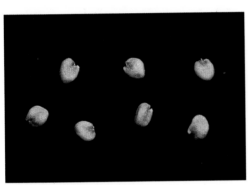

81. *Rorippa palustris*
Bog yellow cress
9.6x

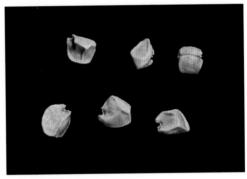

82. *Rorippa sinuata*
Spreading yellow cress
9.6x

83. *Sisymbrium altissimum*
Tumbling mustard
9.6x

84. *Sisymbrium loeselii*
Tall hedge mustard
9.6x

10   20   30

85. *Sisymbrium officinale*
Hedge mustard
9.6x

86. *Thlaspi arvense*
Pennycress
9.6x

87. *Desmanthus illinoensis*
Illinois bundleflower
4.8x

88. *Cassia chamaecrista*
Showy partridge pea
4.8x

89. *Hoffmanseggia glauca*
Indian rush-pea
4.8x

90. *Amorpha canescens*
Lead plant
4.8x

10   20   30

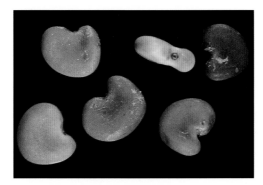

91. Astragalus canadensis
Canada milk-vetch
9.6x

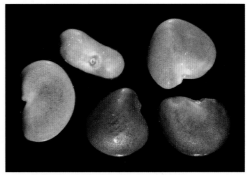

92. Astragalus mollissimus
Woolly locoweed
9.6x

93. Coronilla varia
Crown vetch
4.8x

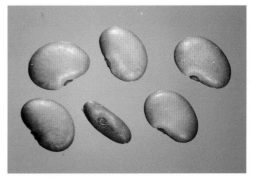

94. Desmodium illinoense
Illinois tickclover
4.8x

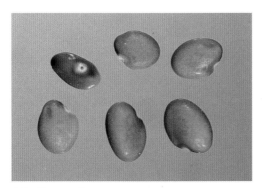

95. Lespedeza cuneata
Sericea lespedeza
9.6x

96. Lespedeza stipulacea
Korean lespedeza
9.6x

10   20   30

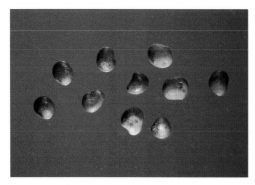

97. *Lotus corniculatus*
Bird's-foot trefoil
4.8x

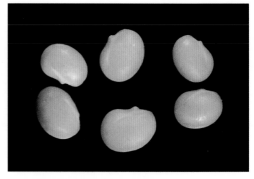

98. *Medicago lupulina*
Black medick
9.6x

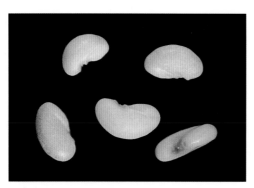

99. *Medicago minima*
Small bur-clover
9.6x

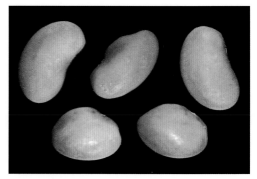

100. *Medicago sativa*
Alfalfa
9.6x

101. *Melilotus alba*
White sweet clover
9.6x

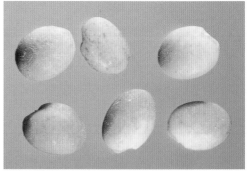

102. *Melilotus officinalis*
Yellow sweet clover
9.6x

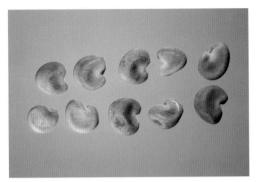

103. *Oxytropis lambertii*
Purple locoweed
4.8x

104. *Strophostyles helvola*
Wild bean
2x

105. *Strophostyles leiosperma*
Slick seed bean
2x

106. *Trifolium campestre*
Low hop-clover
9.6x

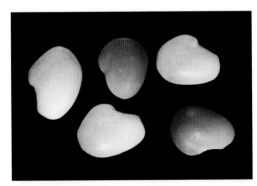

107. *Trifolium pratense*
Red clover
9.6x

108. *Trifolium repens*
White clover
9.6x

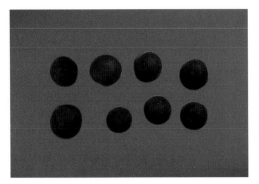

109. *Vicia villosa*
Hairy vetch
2x

110. *Gaura longiflora*
Large-flowered gaura
2x

111. *Gaura parviflora*
Velvety gaura
2x

112. *Oenothera biennis*
Common evening primrose
9.6x

113. *Oenothera laciniata*
Cut-leaved evening primrose
9.6x

114. *Acalypha monococca*
Slender copperleaf
9.6x

10   20   30

115. *Acalypha ostryaefolia*
Hop-hornbeam copperleaf
9.6x

116. *Acalypha rhomboidea*
Rhombic copperleaf
9.6x

117. *Acalypha virginica*
Virginia copperleaf
9.6x

118. *Croton capitatus*
Woolly croton
2x

119. *Croton glandulosus*
Tropic croton
2x

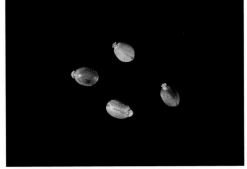

120. *Croton monanthogynus*
One-seeded croton
2x

121. *Croton texensis*
Texas croton
2x

122. *Euphorbia corollata*
Flowering spurge
4.8x

123. *Euphorbia cyathophora*
Fire-on-the-mountain
4.8x

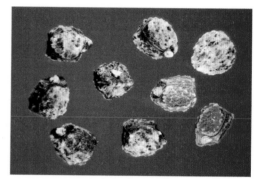

124. *Euphorbia dentata*
Toothed spurge
4.8x

125. *Euphorbia esula*
Leafy spurge
4.8x

126. *Euphorbia maculata*
Spotted spurge
9.6x

127. *Euphorbia marginata*
Snow-on-the-mountain
4.8x

128. *Euphorbia nutans*
Eyebane
9.6x

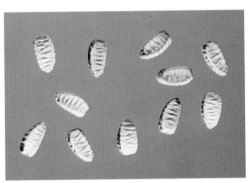

129. *Euphorbia prostrata*
Creeping spurge
9.6x

130. *Tragia betonicifolia*
Nettleleaf noseburn
4.8x

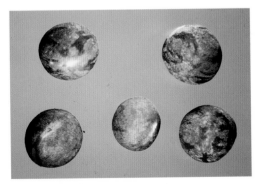

131. *Tragia ramosa*
Noseburn
4.8x

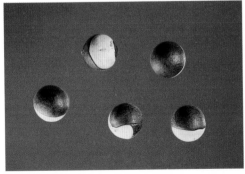

132. *Cardiospermum halicacabum*
Common balloon vine
2x

133. *Rhus glabra*
Smooth sumac
4.8x

134. *Tribulus terrestris*
Puncture vine
2x

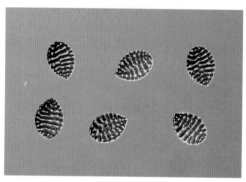

135. *Oxalis dillenii*
Gray-green wood sorrel
9.6x

136. *Oxalis stricta*
Yellow wood sorrel
9.6x

137. *Geranium carolinianum*
Carolina cranesbill
4.8x

138. *Cicuta maculata*
Water hemlock
4.8x

139. *Conium maculatum*
Poison hemlock
4.8x

140. *Daucus carota*
Queen Anne's lace
4.8x

141. *Asclepias subverticillata*
Poison milkweed
2x

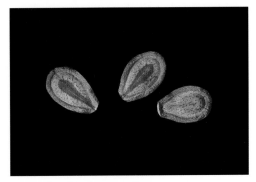

142. *Asclepias syriaca*
Common milkweed
2x

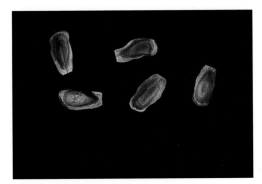

143. *Asclepias verticillata*
Whorled milkweed
2x

144. *Cynanchum laeve*
Sand vine
2x

10   20   30

145. _Datura stramonium_
Jimson weed
4.8x

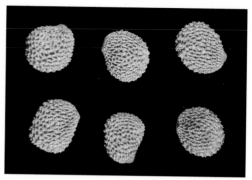

146. _Hyoscyamus niger_
Henbane
9.6x

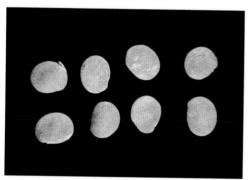

147. _Physalis heterophylla_
Clammy ground cherry
4.8x

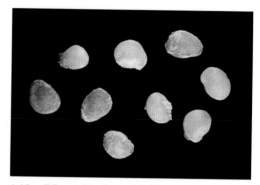

148. _Physalis longifolia_
Common ground cherry
4.8x

149. _Solanum carolinense_
Carolina horse nettle
4.8x

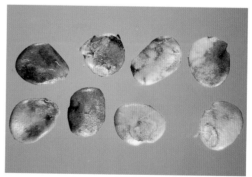

150. _Solanum eleagnifolium_
Silver-leaf nightshade
4.8x

10   20   30

151. *Solanum ptycanthum*
Black nightshade
4.8x

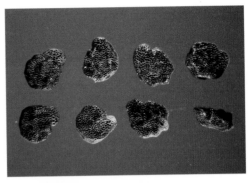

152. *Solanum rostratum*
Buffalo bur
4.8x

153. *Solanum triflorum*
Cut-leaved nightshade
4.8x

154. *Calystegia sepium*
Hedge bindweed
4.8x

155. *Convolvulus arvensis*
Field bindweed
4.8x

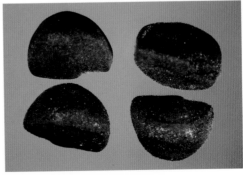

156. *Ipomoea hederacea*
Ivyleaf morning-glory
4.8x

10   20   30

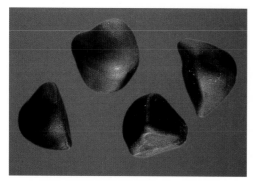

157. *Ipomoea lacunosa*
White morning-glory
4.8x

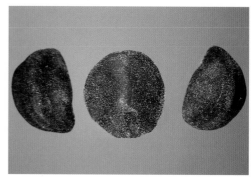

158. *Ipomoea purpurea*
Common morning-glory
4.8x

159. *Cuscuta indecora*
Large alfalfa dodder
9.6x

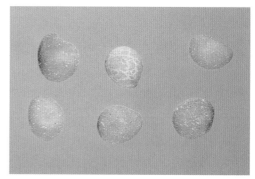

160. *Cuscuta pentagona*
Field dodder
9.6x

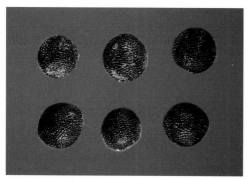

161. *Ellisia nyctelea*
Waterpod
4.8x

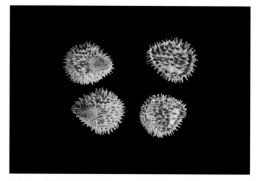

162. *Cynoglossum officinale*
Hound's tongue
2x

163. *Lithospermum arvense*
Corn gromwell
4.8x

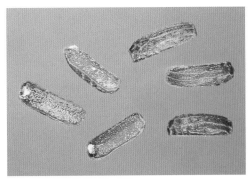

164. *Verbena bracteata*
Prostrate vervain
9.6x

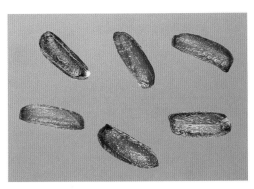

165. *Verbena hastata*
Blue vervain
9.6x

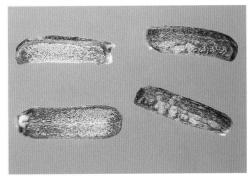

166. *Verbena stricta*
Hoary vervain
9.6x

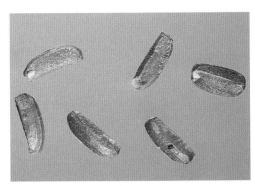

167. *Verbena urticifolia*
Nettle-leaved vervain
9.6x

168. *Glecoma hederacea*
Ground ivy
9.6x

169. *Hedeoma hispidum*
Rough false pennyroyal
9.6x

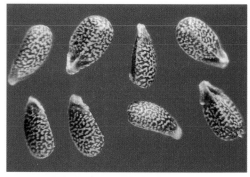

170. *Lamium amplexicaule*
Henbit
9.6x

171. *Salvia reflexa*
Lance-leaved sage
9.6x

172. *Plantago lanceolata*
Buckhorn plantain
9.6x

173. *Plantago major*
Common plantain
9.6x

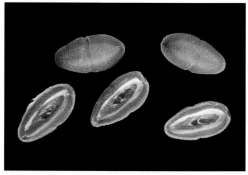

174. *Plantago patagonica*
Patagonian plantain
9.6x

175. *Plantago rugelii*
Rugel's plantain
9.6x

176. *Plantago virginica*
Pale-seeded plantain
9.6x

177. *Verbascum blattaria*
Moth mullein
9.6x

178. *Verbascum thapsus*
Common mullein
9.6x

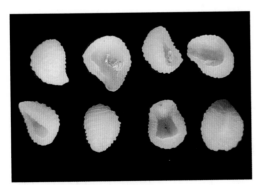

179. *Veronica agrestis*
Field speedwell
9.6x

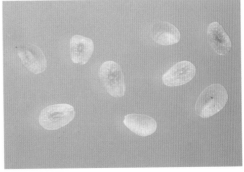

180. *Veronica arvensis*
Corn speedwell
9.6x

10   20   30

181. *Veronica peregrina*
Purslane speedwell
9.6x

182. *Proboscidea louisianica*
Devil's claw
2x

183. *Diodia teres*
Rough buttonweed
2x

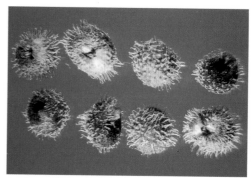

184. *Galium aparine*
Catchweed bedstraw
4.8x

185. *Hedyotis nigricans*
Narrow-leaved bluets
9.6x

186. *Symphoricarpos orbiculatus*
Coral berry
4.8x

10   20   30

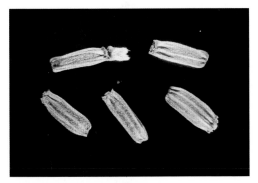

187. *Dipsacus fullonum*
Common teasel
4.8x

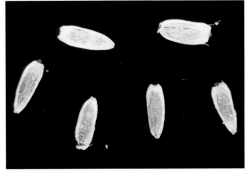

188. *Achillea millefolium*
Yarrow
9.6x

189. *Ambrosia artemisiifolia*
Common ragweed
4.8x

190. *Ambrosia grayi*
Bur ragweed
4.8x

191. *Ambrosia psilostachya*
Western ragweed
4.8x

192. *Ambrosia trifida*
Giant ragweed
4.8x

193. *Arctium minus*
Common burdock
2x

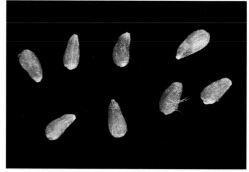

194. *Artemisia absinthium*
Wormwood
9.6x

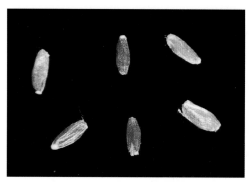

195. *Artemisia ludoviciana*
White sage
9.6x

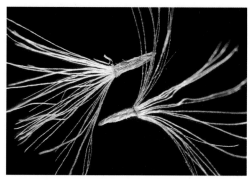

196. *Aster ericoides*
Heath aster
9.6x

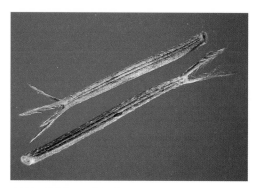

197. *Bidens bipinnata*
Spanish needles
4.8x

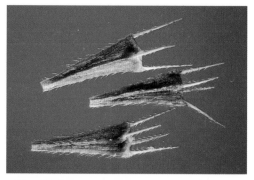

198. *Bidens cernua*
Nodding beggar-ticks
4.8x

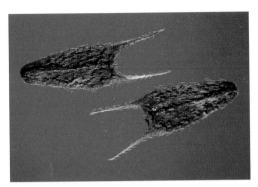

199. *Bidens frondosa*
Devil's beggar-ticks
4.8x

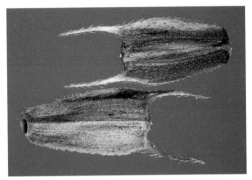

200. *Bidens vulgata*
Tall beggar-ticks
4.8x

201. *Carduus acanthoides*
Plumeless thistle
4.8x

202. *Carduus nutans*
Musk thistle
4.8x

203. *Centaurea cyanus*
Cornflower
4.8x

204. *Centaurea maculosa*
Spotted knapweed
4.8x

10   20   30

205. *Centaurea repens*
Russian knapweed
4.8x

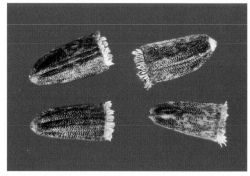

206. *Cichorium intybus*
Chicory
9.6x

207. *Cirsium altissimum*
Tall thistle
4.8x

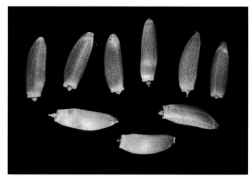

208. *Cirsium arvense*
Canada thistle
4.8x

209. *Cirsium flodmanii*
Flodman's thistle
4.8x

210. *Cirsium ochrocentrum*
Yellow-spine thistle
4.8x

10   20   30

211. *Cirsium undulatum*
Wavyleaf thistle
4.8x

212. *Cirsium vulgare*
Bull thistle
4.8x

213. *Conyza canadensis*
Horseweed
9.6x

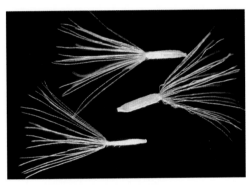

214. *Conyza ramosissima*
Spreading fleabane
9.6x

215. *Coreopsis tinctoria*
Plains coreopsis
9.6x

216. *Dyssodia papposa*
Fetid marigold
4.8x

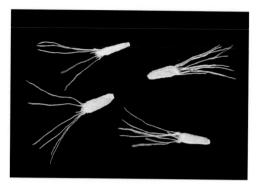

217. *Erigeron strigosus*
Daisy fleabane
9.6x

218. *Eupatorium altissimum*
Tall eupatorium
4.8x

219. *Eupatorium rugosum*
White snakeroot
4.8x

220. *Grindelia squarrosa*
Curly top gumweed
9.6x

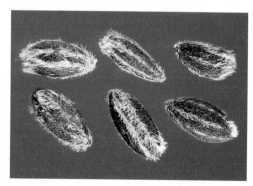

221. *Gutierrezia dracunculoides*
Broomweed
9.6x

222. *Helenium amarum*
Bitter sneezeweed
4.8x

10   20   30

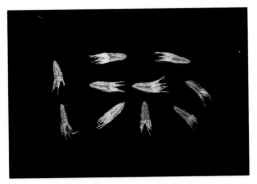

223. _Helenium autumnale_
Sneezeweed
4.8x

224. _Helianthus annuus_
Common sunflower
4.8x

225. _Helianthus ciliaris_
Texas blueweed
4.8x

226. _Helianthus petiolaris_
Plains sunflower
4.8x

227. _Heliopsis helianthoides_
False sunflower
4.8x

228. _Heterotheca latifolia_
Camphor weed
4.8x

229. *Iva annua*
Sump weed
4.8x

230. *Iva axillaris*
Poverty weed
4.8x

231. *Iva xanthifolia*
Marsh elder
4.8x

232. *Kuhnia eupatorioides*
False boneset
4.8x

233. *Lactuca canadensis*
Wild lettuce
4.8x

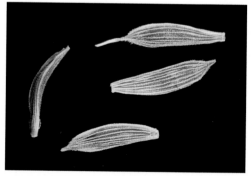

234. *Lactuca saligna*
Willow-leaved lettuce
9.6x

10   20   30

235. *Lactuca serriola*
Prickly lettuce
9.6x

236. *Matricaria matricarioides*
Pineapple weed
9.6x

237. *Onopordum acanthium*
Scotch thistle
2x

238. *Silphium integrifolium*
Whole-leaf rosin-weed
2x

239. *Silphium laciniatum*
Compass plant
2x

240. *Silphium perfoliatum*
Cup plant
2x

241. *Solidago gigantea*
Late goldenrod
4.8x

242. *Solidago rigida*
Stiff goldenrod
4.8x

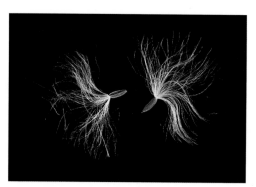

243. *Sonchus arvensis*
Field sow thistle
2x

244. *Sonchus asper*
Prickly sow thistle
2x

245. *Taraxacum laevigatum*
Red-seeded dandelion
4.8x

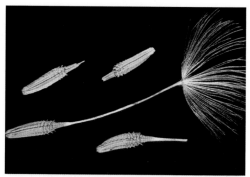

246. *Taraxacum officinale*
Common dandelion
4.8x

|||||||||||||||||||||||||||||||
10   20   30

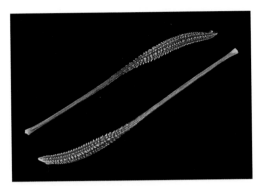

247. *Tragopogon dubius*
Goat's beard
2x

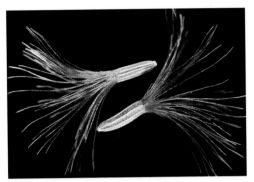

248. *Vernonia baldwinii*
Western ironweed
4.8x

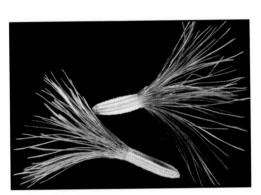

249. *Vernonia fasciculata*
Ironweed
4.8x

250. *Xanthium spinosum*
Spiny cocklebur
2x

251. *Xanthium strumarium* L.
Cocklebur
2x

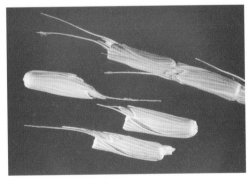

252. *Aegilops cylindrica*
Jointed goat grass
2x

253. *Agropyron repens*
Quackgrass
2x

254. *Agropyron smithii*
Western wheat grass
2x

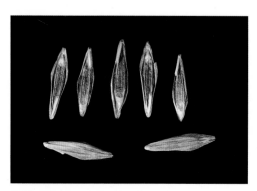

255. *Bromus inermis*
Smooth brome
2x

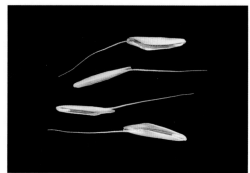

256. *Bromus japonicus*
Japanese brome
2x

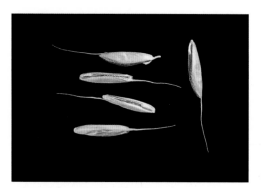

257. *Bromus secalinus*
Cheat
2x

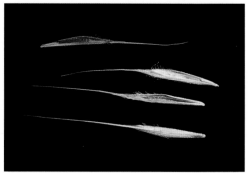

258. *Bromus tectorum*
Downy brome
2x

259. Chloris verticillata
Windmill grass
9.6x

260. Cynodon dactylon
Bermuda grass
9.6x

261. Digitaria ciliaris
Southern crabgrass
9.6x

262. Digitaria ischaemum
Smooth crabgrass
9.6x

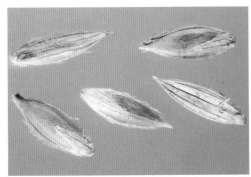

263. Digitaria sanguinalis
Hairy crabgrass
9.6x

264. Echinochloa crus-galli
Barnyard grass
2x

10   20   30

265. *Eleusine indica*
Goosegrass
9.6x

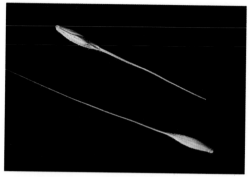

266. *Elymus canadensis*
Canada wild rye
2x

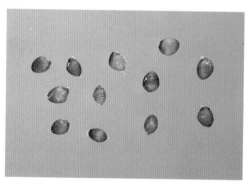

267. *Eragrostis cilianensis*
Stinkgrass
9.6x

268. *Festuca octoflora*
Six-weeks fescue
4.8x

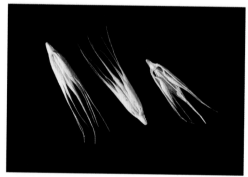

269. *Hordeum pusillum*
Little barley
2x

270. *Panicum capillare*
Common witchgrass
9.6x

271. *Panicum dichotomiflorum*
Fall panicum
9.6x

272. *Panicum miliaceum*
Broom-corn millet
4.8x

273. *Panicum virgatum*
Switchgrass
9.6x

274. *Setaria faberi*
Giant foxtail
9.6x

275. *Setaria glauca*
Yellow foxtail
9.6x

276. *Setaria verticillata*
Bristly foxtail
9.6x

10   20   30

277. *Setaria viridis*
Green foxtail
9.6x

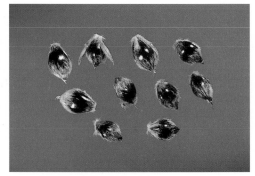

278. *Sorghum bicolor*
Shattercane
2x

279. *Sorghum halepense*
Johnson-grass
2x

280. *Sporobolus neglectus*
Poverty grass
9.6x

Length 0.9–1.2 mm; width 0.6–0.7 mm; thickness ±0.3 mm.
Lawns; moist soil, open areas; central and SE 1/4 GP.

181. *Veronica peregrina* L.   Purslane speedwell

Seed: Outline elliptical. Cross section a very narrow ellipse with a small knob at the center of one long edge. Form like a very thin chip, variously bent. One face has a central ridge, extending from one end most of the length of the seed. The hilum is on this ridge.

Surface smooth; shiny. Light orange-brown, translucent; the ridge is red-brown.

Length 0.8–0.9 mm; width 0.4–0.5 mm; thickness 0.2–0.4 mm.
Cultivated fields, ditches; wet or moist rich soil; throughout GP.

**Pedaliaceae
Unicorn-plant
Family**

182. *Proboscidea louisianica* (P. Mill.) Thell.   Devil's claw

Seed: Outline obovate with ragged edges. Cross section narrowly ovate and ragged. Form obovoid, compressed; one face is nearly flat, and the other face has a ridge that extends from the small end of the seed to the center. The seed is usually thicker toward the wide end. The hilum is a cavity in the end of the ridge, near the base of the seed.

Surface very rough, with irregular angular knobs; glittery when seen in strong light. There is also a fine overall reticulate pattern, visible at 7× magnification. Black.

Length 9.0–9.5 mm; width 5.4–6.0 mm; thickness 1.5–2.0 mm.
Corn, sorghum, and soybean fields, waste places, barnyards; often on sandy soil; S 1/2 GP.

**Rubiaceae
Madder Family**

183. *Diodia teres* Walt.   Rough buttonweed

Nutlet: Outline rounded oblong to obovate. Cross section plano-convex. Form like half of an obovoid shape. Dorsal surface convex; ventral surface flat. Thicker at the apex than base. On the flat surface is a large, low, U-shaped ridge and a narrow central lengthwise ridge (a scar from the attachment of one nutlet to another). At the apex of the nutlet there may be 2 or 3 persistent leaflike calyx remnants.

Surface of the convex face scurfy, strigose. Surface of flat face rough and lacerate, lacking hairs. Dark orange-brown; the surface color obscured by a scurfy layer and hairs on the dorsal side, making it appear dull gray-brown.

Length 3.0–3.6 mm; width 2.4–2.9 mm; thickness 1.4–1.8 mm.
Pastures, waste places; dry, rocky or sandy soil; SE 1/4 GP.

184. *Galium aparine* L.   Catchweed bedstraw

Nutlet: Outline round to short-elliptical. Cross section short-ellipsoid with a shallow notch in one long side. Form spheroid or ellipsoid, with a single deep depression. This pit is ±0.5 mm in diameter, covered with thin scarlike tissue. At one end of the pit the surface is elevated and has a distinct attachment point. The nutlet is hollow; the wall is ±0.5 mm thick.

Surface closely covered with hooked bristles, ±0.6 mm long; the surface between the bristles is rough. Gray-brown overall; the surface is dark brown, and the bristles are transparent, colorless, or orange-flecked.

Length 1.9–3.0 mm; width 1.6–2.5 mm; thickness 1.4–2.3 mm. (Measurements exclude bristles.)

Gardens, thickets; shaded areas or moist soil; throughout GP.

185. *Hedyotis nigricans* (Lam.) Fosb.   Narrow-leaved bluets

Seed: Outline a rounded oblong. Cross section oblong or irregular. Form oblong, rounded, compressed, with thin margins. The seed is usually bent and generally irregular and wrinkled. There is a thin marginal wing around one small end. The hilum is near the center of one face.

Surface finely textured; dull. At 15× it appears shiny, and a fine alveolate pattern is visible. Black.

Length 1.1–1.2 mm; width 0.5–0.6 mm; thickness 0.3 mm.

Pastures; dry, rocky soil; S 1/2 GP.

Caprifoliaceae Honeysuckle Family

186. *Symphoricarpos orbiculatus* Moench   Coral berry, buckbrush

Seed: Outline elliptical. Cross section elliptical with 1 straight edge. Form ellipsoid, compressed, with 1 flat face and 1 convex face. There is a small point at each end.

Surface smooth; dull. Under high magnification it appears finely textured. Pale tan.

Length 3.4–3.9 mm; width 2.0–2.2 mm; thickness 1.1–1.2 mm.

Rangeland, pastures, fence rows; open areas; SE 1/4 GP.

Dipsacaceae Teasel Family

187. *Dipsacus fullonum* L.   Common teasel

Achene: Outline nearly rectangular. Cross section square. Form like a long box with a ridge on each long angle. There is usually a lengthwise rib on the center of each long face. The style base extends slightly beyond the depressed apex; the ciliate calyx, ±1 mm long, may be persistent on the apex of the achene. The attachment point is at the center of the base.

Surface conspicuously ribbed; dull. There are very fine straight white hairs over the entire surface; these are visible at 10× magnification. Light brown, with pale ridges.

Length 3.1–3.7 mm; thickness 0.8–1.1 mm.

Old fields, pastures, roadside ditches, often found as an escape from cultivation; open areas; E 1/4 GP.

Asteraceae
Sunflower Family

188. *Achillea millefolium* L.   Yarrow

Achene: Outline long obovate. Cross section narrowly elliptical. Form obovoid, compressed. The seed itself is obovate; it is enclosed in a larger membranous pericarp that has a marginal wing. There is a small round rim around the basal attachment point, a short apical collar, and a short beak. There is no pappus.

Surface membranous, slightly striate; slight sheen. The pericarp is white, but the seed is dark brown and visible through the pericarp.

Length 1.6–1.8 mm; width 0.4–0.5 mm; thickness 0.2–0.3 mm.

Pastures, old fields, fence rows; disturbed soil; throughout GP.

189. *Ambrosia artemisiifolia* L.   **Common ragweed**

Bur (woody structure derived from the involucre; encloses a single achene): Outline obovate with an extended beak. Cross section round or somewhat 3-sided. Form like a top. The beak, projecting from the center of the broad end, is ±1 mm long and 0.5 mm broad at its base. Under high magnification several teeth are visible around an opening at the tip of the beak. From the widest part of the bur a series of 6 to 8 pointed tubercles projects upward. These tubercles are ±0.3 mm long. There is an oblique, round attachment scar at the base of the bur.

Surface striate; dull. There are scattered white granules, mostly on the upper half. Fine white strigose hairs are present above the tubercles and on the beak. Dark gray-brown.

Length 3.5–3.9 mm; thickness 1.6–2.0 mm.

(The burs are often found in varying stages of deterioration. The tubercles are easily worn off. The achene itself is obovate with a small beak, smooth, light brown; length ±3.1 mm, beak ±0.4 mm long.)

Rangeland, waste areas, cultivated fields, old pastures; disturbed soil; throughout GP except SW 1/8.

190. *Ambrosia grayi* (A. Nels.) Shinners

Bur ragweed, woolly bursage

Bur (woody, spiny structure derived from the involucre; encloses 2 achenes): Outline obovate with stout spreading spines and 2 incurved teeth at the apex. Cross section round or slightly compressed, very spiny. Form obovoid, with a narrow blunt base. At the apex there are 2 inwardly curved fine-tipped beaks, ±1 mm long. On the body of the bur are about 12 to 14 spreading spines, ±1 mm long. The spines are narrower than the beaks and are usually hooked at the tip.

Surface tough, woody; dull. It is densely covered with transparent yellow resinous droplets visible at 10× magnification. There are scattered long white hairs. Pale olive.

Length 4.0–6.5 mm; diameter 2.8–3.7 mm. (Measurements exclude spines and beaks.)

The achene is long obovate in outline, plano-convex in cross section. The surface is striate and streaked with black and dark brown. Length ±4.1 mm; width ±1.8 mm; thickness ±1.2 mm.

Cultivated fields, fallow fields; moist, often saline soil; SW 1/4 and central GP.

191. *Ambrosia psilostachya* DC.

Western ragweed

Bur (woody structure derived from the involucre; encloses a single achene): Outline obovate with a small beak. Cross section round to rounded triangular. Form ovoid, irregular, with a narrow, oblique base from which a rounded keel runs halfway up the side and with a small apical beak. Under magnification the beak appears papery, with several small teeth at the tip. It is ±0.5 mm long and ±0.3 mm broad at its base. Small tubercles may project upward from the widest area of the bur. There are 1 or 2 of these, or sometimes none; rudimentary tubercles may be visible with magnification. The tubercles are up to 0.3 mm long.

Surface papery with a few irregular wrinkles. There are white strigose hairs; these are more abundant on the upper half of the bur and on the beak. Mottled dark and light gray-brown.

Length 3.2–3.5 mm; thickness 1.7–2.0 mm.

The achene is obovate with a slight beak, smooth, very dark brown.

Pastures, rangeland; open areas; throughout GP.

192. *Ambrosia trifida* L.   Giant ragweed

Bur (woody structure derived from the involucre; encloses a single achene): Outline obovate with an extended beak. Cross section rounded and 5-sided. Form like a top, somewhat 5-sided, with a small blunt base and a narrow cuspidate beak. Five lengthwise rounded angles extend from the base to just above the broadest part of the bur. Each of these angles terminates in a small acute vertical tooth. Above this ring of 5 teeth there may be a second series of 1, 2, or 3 similar teeth, alternate with the first series. The basal attachment scar is round.

Surface (lower portion of the bur) smooth, papery, with fine striations. The upper portion and the beak are wrinkled and have white strigose hairs, which are more abundant on the beak. Greenish tan; there are deep purple streaks and mottles on the upper portion and between the angles on the lower part.

Length 5.6–6.3 mm; thickness 3.0–3.8 mm.

The achene is obovate, with a very small apical beak, smooth. Gray-brown; under magnification black and brown patches are evident.

Roadsides; moist or rich disturbed soil; throughout GP.

193. *Arctium minus* Bernh.   Common burdock

Achene: Outline clavate. Cross section dish shaped. Form clavate, very flattened, often curved to one side; one broad face is slightly convex, the other slightly concave. Both the base and the apex are truncate. The attachment scar, at the base, is ±0.8 mm in diameter. There is a slight apical rim, ±1 mm in diameter, and a very short beak. The readily deciduous pappus consists of barbed yellow bristles, ±1.5 mm long.

Surface finely textured; slight sheen. There is usually a narrow central ridge on each broad face. With high magnification, many small lengthwise wrinkles or tubercles are visible. Mottled brown and very dark brown.

Length 5.7–6.1 mm; width 2.2–2.4 mm; thickness 0.8–1.0 mm.

Pastures, feedlots; partially shaded areas, moist soil; throughout GP except SW 1/4.

194. *Artemisia absinthium* L.   Wormwood

Achene: Outline clavate. Cross section elliptical. Form clavate, somewhat compressed, often appearing curved or bent in edge view. The basal attachment point is small, round. At the center of the apex is a small round scar, but there is no apical collar and no pappus.

Surface striate; silky sheen. Yellow-brown, somewhat translucent. Length 1.0–1.1 mm; width 0.4–0.6 mm; thickness 0.3–0.4 mm. *Roadsides; open areas; N 1/4 GP.*

195. *Artemisia ludoviciana* Nutt. White sage

Achene: Outline oblong to elliptical. Cross section 5-sided. Form 5-sided, elongate, tapered slightly to both the base and the apex. The whole achene is often slightly curved. There is a small apical collar.

Surface distinctly striate; silky luster. Yellow-brown, somewhat translucent.

Length 1.0–1.2 mm; thickness 0.4–0.5 mm. *Pastures, rangeland, waste places; dry open areas; throughout GP.*

196. *Aster ericoides* L. Heath aster, white aster

Achene: Outline oblong but narrowed toward the base. Cross section very thin. Form oblong, compressed, very thin, with one or both long edges curved inward to the narrow base. There is a small thin apical collar and a pappus of many minutely barbed capillary hairs, ±3.6 mm long.

Surface covered by many short white appressed hairs that may give the achene a shiny appearance. Light brown; the pappus is tan. Length 1.0–1.3 mm; width 0.3–0.4 mm; thickness ±0.2 mm. *Pastures, rangeland, roadsides; open areas; throughout GP.*

197. *Bidens bipinnata* L. Spanish needles

Achene: Outline a very narrow ellipse with 2, 3, or 4 awns flaring from the apex. Cross section rhombic. Form 4-sided, very long, narrowed slightly to both the base and the apex. The base is obliquely truncate. The attachment scar, in the truncate area, is large and has a smooth circular rim. The pappus of 2, 3, or 4 stout awns, 1.5–3.5 mm long, arises from the apex. The awns are retrorsely barbed with stiff transparent spines that are ±0.5 mm long. There are about 8 lengthwise ridges on the body of the achene: 1 on each of the 4 lengthwise angles and usually 1 on the center of each face.

Surface smooth; dull. At high magnification it appears finely papillate. Dull brown, appearing mottled at high magnification; the truncate area and the awns are pale.

Length (0.8) 1.1–1.6 cm; width 0.9–1.2 mm; thickness 0.8–1.0 mm. (Measurements exclude the awns.)

Cultivated fields, waste places, gardens; rich soil, partial shade; S 1/2 GP except SW 1/8.

198. *Bidens cernua* L.   Nodding beggar-ticks

Achene: Outline obtriangular, with 4 prominent awns spreading from the apex. Cross section usually narrowly triangular. Form elongate, flaring toward the apex, compressed. One side is nearly flat and has a central ridge; the other side is composed of 2 faces that meet in a central ridged angle. The apex is convex and rhombic and has a small beak. Each of the side margins and central ridges is prolonged into a sturdy awn, ±1.9–3.0 mm long. The ridges, margins, and awns all have retrorse barbs, ±1.5 mm long. The awns arising from the side margins have 3 rows of barbs; the awns arising from the center margins have 2 rows of barbs.

Surface striate; somewhat shiny at 7× magnification. There are scattered tubercles with retrorse barbs arising from them. The apical area is shiny. Dark brown; apex straw colored; awns pale, with very pale to dark barbs.

Length 5.2–6.7 mm; width 1.1–1.6 mm; thickness 0.5–0.7 mm. (Measurements exclude awns.)

Roadside ditches; wet or moist disturbed soil; throughout GP.

199. *Bidens frondosa* L.   Devil's beggar-ticks

Achene: Outline obovate with 2 prominent awns spreading from the apex. Cross section very thin, concave-convex or biconvex. Form oblong, very thin, slightly bent. There is a lengthwise ridge in the center of each face. The base is truncate with a smooth rim. The 2 rigid awns of the pappus, about 2.0–4.5 mm long, are retrorsely barbed. The apex is concave with the concavity extending up the inner surface of each awn. A small remnant of the style may be present at the center. There may be a small tooth on the rim of the concave area at the center of each face.

Surface scabrous; very dull. It appears warty under magnification. At high magnification the entire surface appears finely striate. Sparse short straight brownish hairs arise from the midribs and warts. The awns have short stout retrorse barbs in 3 rows. Dark brown to brownish black; the awns are paler.

Length 4.7–7.3 mm; width 1.5–2.8 mm; thickness 0.5–0.8 mm.

Pastures, cultivated fields, roadside ditches; moist soil; through-out GP.

200. *Bidens vulgata* Greene   Tall beggar-ticks

Achene: Outline oblong but flared toward the apex, with 2 prominent awns spreading from the apex. Cross section very thin, slightly curved. Form oblong, very compressed, with the dorsal face convex and the ventral face concave. The basal attachment area has an elliptical rim, ±1.2 mm long. On each face there is a central lengthwise ridge. The pappus of 2 sturdy curved awns arises from either side of the apex. There is a row of small barbs on each side margin of the achene. The awns each have 3 rows of sturdy retrorse barbs.

Surface striate, with scattered tubercles; dull. There are scattered short straight brown hairs, ±0.2 mm long. Brown; the awns are paler, and the barbs are transparent.

Length 7.4–8.9 mm; width 3.2–3.5 mm; thickness ±0.6 mm. (Measurements exclude the awns.)

Ditches; moist disturbed soil; throughout GP except S 1/4.

201. *Carduus acanthoides* L.   Plumeless thistle

Achene: Outline elliptical, often curved to one side. Cross section variable, often elliptical. Form ellipsoid, compressed. Basal attachment point small, oblique. Apex truncate; apical collar thin, slightly flaring, ±0.2 mm long. Apical beak stout, conspicuous, ±0.4 mm in diameter, extending ±0.3 mm beyond collar. Pappus deciduous.

Surface appears slightly wrinkled or tuberculate at 10× magnification; shiny. Tan with about 8 faint, narrow brown stripes on each face; collar pale.

Length 2.8–3.0 mm; width 1.2–1.3 mm; thickness 0.5–0.7 mm.

Pastures, roadsides; open areas; NE 1/8, NW 1/4, central, and E central GP.

202. *Carduus nutans* L.   Musk thistle, nodding thistle

Achene: Outline elliptical, curved. Cross section variable, somewhat elliptical. Form ellipsoid, compressed, curved; sometimes faintly 4- or 5-sided. The attachment point is a 5-sided hollow at the base. The apex is obliquely truncate and has a distinct thin, slightly flaring collar. There is a large conical style base at the center of the apex. This beak is ±0.7 mm broad at the base and projects 0.5 mm beyond the collar.

The pappus, which is readily shed, consists of many fine bristles arising from the collar. The pappus bristles are straight and tannish white,
about 15 mm long. At high magnification many fine, very short barbs
are visible on the bristles.

Surface smooth; very shiny. There are many slight lengthwise
ridges. Straw colored; the style base is yellow-brown or orange, and
with magnification dark, faint, lengthwise dotted stripes can be seen
on the body of the achene.

Length (including beak) 3.4–4.5 mm; width 1.3–1.9 mm; thickness
0.8–1.3 mm.

*Roadsides, pastures, rangeland, waste places; open areas; central and
E central GP.*

203. *Centaurea cyanus* L.   Cornflower

Achene: Outline oblong but somewhat tapered to the base and with a
large notch missing from one side near the base; there is a pappus of
bristles about as long as the achene. Cross section elliptical. Form oblong, compressed, with a narrow base and truncate apex. Along one
margin, near the basal end, is a conspicuous notch, ±1.2–1.5 mm long.
There is a deep cavity in this notch at the attachment point. At the apical end is a thin, slightly flaring collar, ±0.2 mm long. The pappus of
many broad bristles arises from within the collar. The bristles are
short plumose and vary from ±0.4 to 3.0 mm in length.

Surface smooth, shiny, with faint narrow lengthwise ridges. It is
covered with many fine short white hairs that are easily worn off.
There is also a tuft of white hairs at the base of the achene. Glaucous
pale reddish brown; the collar and notch area are hairless and straw
colored. The pappus is dull orange to purplish.

Length (excluding pappus) 3.2–4.0 mm; width 1.2–1.8 mm; thickness 0.9–1.3 mm.

Roadsides, wheat fields; good soils; E 2/3 GP.

204. *Centaurea maculosa* Lam.   Spotted knapweed

Achene: Outline generally oblong, with curved long edges; there is a
large shallow notch on one long edge near the base and a conspicuous
pappus that is about 2/3 as long as the achene. Cross section ovate.
Form ellipsoid with a truncate apex, compressed. There is a large
notch in the side margin near the base. The notch is partly filled with
corky tissue. There is a slightly flaring apical collar, ±0.1 mm long.
The persistent pappus arises within the collar. The pappus bristles are

sturdy, short plumose or pectinate, white, and of mixed lengths, varying from ±0.2 to 2.0 mm.

Surface striate and with about 10–15 mostly faint ridges; somewhat lustrous. There are sparse fine white hairs, ±0.2 mm long, over the body of the achene. Gray-brown; collar and ridges pale yellow, and the basal area smooth, shiny, and pale.

Length 2.8–3.3 mm; width 1.2–1.6 mm; thickness 0.9–1.1 mm. (Measurements exclude the pappus.)

Pastures, roadsides, cultivated fields; good soils; throughout GP.

205. *Centaurea repens* L. Russian knapweed

Achene: Outline clavate to elliptical but tapered to a narrow base. Cross section a narrow oblong. Form oblong, compressed, very thin, narrowed to the base and truncate at the apex. The deciduous pappus is a ring of bristles arising from the apex. These bristles are white to straw colored, slightly flattened and short plumose. They are of mixed lengths, from ±0.3 to 1.3 mm.

Surface covered with many fine soft white hairs. There are several lengthwise ridges on each face. Gray-brown.

Length 2.1–2.4 mm; width 0.6–0.7 mm; thickness ±0.3 mm.

Cultivated fields, irrigation ditches, waste places; good soils; N 1/4 and E 1/2 GP.

206. *Cichorium intybus* L. Chicory

Achene: Outline long-triangular with more or less curved sides. Cross section somewhat angular with about 3–5 unequal sides. Form peglike or conical, narrow at the base and truncate at the apex, irregularly curved, ridged, and angled. The pappus is a fringe of membranous rounded teeth, ±0.2 mm long, around the apical end. Within this fringe is a smooth flat ring that has a short style remnant, ±0.1 mm long, in the center. The attachment point is at the base.

Surface smooth; slight sheen. There are about 10 lengthwise ridges. At 10× magnification the surface appears finely textured. At high magnification a fine pattern of crosswise wrinkles is visible. Gray-brown, with pale base and apex; with magnification light and dark brown mottling is apparent. The pappus is off-white.

Length (including the pappus) 2.1–3.0 mm; thickness 0.7–1.4 mm.

Roadsides; dry open areas; E central GP.

207. *Cirsium altissimum* (L.) Spreng.

Tall thistle, pasture thistle

Achene: Outline generally oblong but narrowed toward the base and slightly narrowed toward the apex. Cross section biconvex to rhombic. Form ellipsoid, with 1 straight long edge; the other long edge is either convex or bent inward near the apex. The attachment point is at the base. The apex is truncate at a slight angle. There is a thin apical collar, ±0.1 mm long. A blunt beak, ±0.4 mm in diameter, extends ±0.3 mm beyond the collar. The deciduous pappus of many bristles is white, ±2.7 cm long, and feathery with very fine soft hairs ±2–3 mm long.

Surface finely striate; highly glossy in the apical fifth and slightly glossy in the lower portion. Light yellow in the apical fifth, light brownish straw colored in the lower portion.

Length including beak (4.6) 4.9–5.1 mm; width 1.8–1.9 mm; thickness 0.8–1.0 mm.

Pastures, ditches, roadsides, waste areas; good soils; E 1/2 GP.

208. *Cirsium arvense* (L.) Scop.

Canada thistle

Achene: Outline elliptical, with a truncate apex. Cross section rounded and often 3- or 4-sided. Form ellipsoid, slightly compressed, often 4-sided; the whole achene is often curved to one side. The base is small and blunt. The apex is truncate and has a thin apical collar, ±0.2 mm long. The beak extends ±0.2 mm beyond the collar. The deciduous pappus arises from the collar. It consists of many tan bristles, ±2.8 cm long, that are feathery with many fine soft hairs, ±2 mm long.

Surface finely striate; glossy. Light brown; the collar is yellow.

Length (including the beak) 3.3–4.0 mm; width 1.0–1.2 mm; thickness 0.7–0.9 mm.

Cultivated fields, pastures, waste areas; moist soil; N 2/3 GP.

209. *Cirsium flodmanii* (Rydb.) Arthur

Flodman's thistle

Achene: Outline short elliptical but narrower toward the basal end and slightly flared at the apex; both the base and apex are truncate. Cross section oblong to rhombic. Form ellipsoid, compressed. There is a thin, slightly flaring apical collar, ±0.3 mm long, and an apical beak that extends ±0.2 mm beyond the collar. The deciduous pappus arises from the collar. It consists of many tan bristles, ±3.0 cm long, that are feathery with many fine soft hairs, ±6 mm long.

Surface striate; shiny. Light brown. The collar is smooth, straw colored.

Length 3.9–4.0 mm; width 1.8–2.0 mm; thickness 0.6–1.1 mm.
Rangeland, pastures; moist soil; N 1/2 GP.

210. *Cirsium ochrocentrum* A. Gray   Yellow-spine thistle

Achene: Outline somewhat elliptical with a truncate apex. Cross section elliptical. Form ellipsoid, compressed. The attachment scar is an ovate cavity in the oblique base. The apex is oblique, and there is an apical collar, ±0.2 mm long and ±1.4 mm in diameter. The beak, ±0.4 mm in diameter, projects ±0.3 mm beyond the collar. The pappus, which arises from the collar, is readily shed and tends to fall as a unit. It consists of many tan bristles, ±2.3 cm long, that are feathery with many fine soft hairs, ±1 mm long.

Surface glossy but finely roughened with faint ridges. Light orange-brown with fine red-brown streaks; the collar is yellow.

Length 4.9–5.5 mm; width 2.7–3.1 mm; thickness 1.5–1.8 mm.
Pastures, rangeland; dry open areas; SW 1/4 GP.

211. *Cirsium undulatum* (Nutt.) Spreng.   Wavyleaf thistle

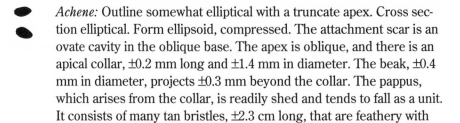

Achene: Outline generally elliptical but tapering to an acute base and obliquely truncate at the apex. Cross section ovoid. Form ellipsoid, compressed, sometimes slightly curved. There is a thin flaring apical collar, ±0.7 mm long, and a conspicuous beak, ±0.6 mm in diameter, that extends ±0.5 mm beyond the collar. The deciduous pappus arises from the collar. It consists of many straight white or slightly tawny bristles, ±3.5 cm long, that are feathery with many fine soft hairs, ±0.3–0.5 mm long.

Surface smooth; glossy. There are faint lengthwise striations, and under magnification the surface appears slightly rugose. Pale straw color; the collar is paler, and there are a few pale narrow lengthwise stripes, with darker mottling between the stripes.

Length 5.3–6.3 mm; width 2.2–2.5 mm; thickness 1.0–1.5 mm.
Pastures, rangeland, roadsides; dry soil; throughout GP.

212. *Cirsium vulgare* (Savi) Ten.   Bull thistle

Achene: Outline generally oblong but narrowed toward the base and swollen near the apex. Cross section variable, somewhat 4-angled. Form ellipsoid, straight or slightly curved. The apex is truncate obliquely or at a right angle. There is a thin apical collar, +0.1 mm long, and an apical beak that is ±0.3 mm in diameter and extends ±0.2–0.3 mm beyond the collar. The deciduous pappus consist of many straight white bristles, ±1.4–2.0 cm long, that are feathery with many very fine soft hairs, ±0.2–0.5 mm long.

Surface nearly smooth, with slight lengthwise striations; glossy. Dull orange-brown to pale gray-brown; the collar is paler, and there are a few dark, narrow lengthwise streaks.

Length 3.2–4.0 mm; width 1.2–1.3 mm; thickness 0.8–1.0 mm.

Pastures, rangeland, waste areas; dry soil; throughout GP except SW 1/4.

213. *Conyza canadensis* (L.) Cronq.   Horseweed

Achene: Outline oblong but narrowed toward the base and with a persistent pappus of hairs extending from the apex. Cross section elliptical. Form oblong, compressed, with thin side margins. There is a smooth circular rim around the basal attachment point. At the apex there is a small collar. The pappus, arising from the collar, is a ring of about 20 minutely barbed hairs, ±2 mm long.

Surface smooth; silky sheen. There are many short appressed white hairs. Pale straw color; the basal rim, collar, and pappus are off-white.

Length 1.0–1.2 mm; width ±0.3 mm; thickness ±0.2 mm. (Measurements exclude the pappus.)

Roadsides, pastures, cultivated fields; disturbed soil; throughout GP.

214. *Conyza ramosissima* Cronq.   Spreading fleabane

Achene: Outline oblong but tapered to a narrow base and with a persistent pappus of hairs spreading from the apex. Cross section a very thin ellipse. Form oblong, compressed, with a narrowed, obliquely truncate base. The 2 long edges are narrowly winged. The attachment point, at the base, is small and round. There is a slight apical collar. The pappus, arising from the collar, consists of minutely barbed hairs, ±2.2 mm long.

Surface covered with many short, straight appressed hairs. The hairs have a soft sheen when seen with magnification. Pale yellow.

Length 1.2–1.3 mm; width 0.3 mm; thickness 0.1–0.2 mm.

Lawns, cultivated fields; open areas, disturbed soil; SE 1/4 GP.

215. *Coreopsis tinctoria* Nutt.   Plains coreopsis

Achene: Outline elliptical, but both base and apex are truncate. Cross section rhombic or broadly wedge shaped. Form ellipsoid, sectorlike, slightly curved around the crosswise axis; the dorsal face is convex and may have a narrow central lengthwise ridge, and the ventral face has a rounded central ridge. The basal attachment area is extended into an elliptical rim. The apical collar is smaller than the basal rim. There is a small beak at the apex. The pericarp is thin and flexible. Pappus very small, deciduous.

Surface striate; dull at 7× magnification. At 10× it appears finely papillate, with the papillae in lengthwise rows; somewhat shiny. Most achenes also have scattered large tubercles on 1 or both faces. The basal attachment area is smooth, shiny. Brown-black to dark brown; attachment area, apical collar, and tubercles light brown.

Length 1.6–2.3 mm; width 0.6–0.9 mm; thickness 0.3–0.4 mm. *Roadside ditches; moist disturbed soil; throughout GP.*

216. *Dyssodia papposa* (Vent.) Hitchc. Fetid marigold

Achene: Outline obtriangular, narrow, with the persistent pappus spreading from the apex. Cross section irregularly 5-sided. Form elongate, 5-sided and 5-angled, often compressed. The basal attachment point is an oblique smooth knob. The persistent pappus, ±3 mm long, arises from the truncate apex. The pappus consists of ±20 thin elongate scales, each with about 10 minutely barbed long bristles.

Surface finely roughened; dull. There are straight white hairs, ±0.2 mm long, over the entire surface or mostly on the angles. Black; pappus straw colored and bristle tips purplish.

Length 3.2–3.8 mm; width 0.7–0.8 mm; thickness 0.3–0.7 mm. *Roadsides; open areas, dry soil; throughout GP except NW 1/8 and SW 1/8.*

217. *Erigeron strigosus* Muhl. ex Willd. Daisy fleabane

Achene: Outline a rounded oblong, slightly wider at the apex. Cross section a thin ellipse or somewhat rhombic. Form oblong, compressed, very thin, with narrow marginal wings on the long edges. There is a slight lengthwise ridge on the center of each face. A smooth circular rim surrounds the basal attachment point. The pappus consists of an inner ring of bristles and an outer ring of membranous teeth. There are about 10 of these bristles, ±1.7 mm long and minutely upwardly barbed; they are easily broken off. The teeth are flaring, ±0.1 mm long, and persistent. There is a very small style remnant at the center of the apex.

Surface finely striate; silky sheen. There are scattered short white appressed hairs. Pale straw color, somewhat translucent.

Length 0.8–0.9 mm; width 0.3 mm; thickness 0.1–0.2 mm.

Roadsides, meadows; disturbed soil; throughout GP except SW 1/8.

218. *Eupatorium altissimum* L.   Tall eupatorium

Achene: Outline oblong. Cross section variable, generally elliptical. Form oblong, terete to compressed. There is an oblique, smooth knob at the basal attachment point. At the apex there is a slightly flaring apical collar, ±0.2 mm long, with a persistent pappus of ±25 white or yellowish bristles, ±5.5 mm long, that are feathery with many fine hairs, ±0.5 mm long.

Surface with about 15 closely spaced narrow ribs; dull. There are scattered minute hairs on the ribs. The hairs are yellowish, upwardly directed. Brown with yellow undertones; collar and basal attachment area yellow.

Length 4.2–5.0 mm long; width 0.6–0.7 mm; thickness 0.2–0.5 mm.

Pastures; disturbed soil; SE 1/4 GP.

219. *Eupatorium rugosum* Houtt.   White snakeroot

Achene: Outline oblong, narrowed toward the base. Cross section usually 5-sided. Form oblong, narrowed toward the base and slightly constricted at the apex; 5-angled, the angles narrowly winged. The deciduous pappus of ±30 tannish-white bristles, ±2.2 mm long, arises from a short apical collar. The pappus bristles are antrorsely barbed; the barbs are visible at 10× magnification.

Surface finely textured; dull at 10× magnification. At 30× magnification it is scalariform, with a slight sheen. Some achenes have a sparse covering of short white hairs, visible at 10×. Black or silvery black; base and apex tan.

Length 1.9–2.6 mm; diameter 0.5–0.6 mm.

Thickets; shaded areas; E 1/3 GP.

220. *Grindelia squarrosa*   Curly top gumweed
(Pursh) Dun. var. *squarrosa*

Achene: Outline roughly oblong, usually curved, with the sides tapered to a blunt-tipped base. Cross section variable, usually rhombic. Form oblong, compressed, often bent, with a narrow base and obliquely truncate apex. The long edges are slightly winged. The attachment point, at the base, is small and round. There is a slight thin apical collar and small style beak, ±0.1 mm long.

Surface smooth, papery; silky luster. There are several fine lengthwise ridges. Pale grayish tan.

Length 2.9–4.0 mm; width 1.6–2.3 mm; thickness 0.8–1.4 mm.
Roadsides, rangeland; dry soil; throughout GP.

221. *Gutierrezia dracunculoides* (DC) Blake — Broomweed

Achene: Outline obovate. Cross section round to 3-sided. Form obovoid, tapered to a narrow point at the base; sometimes slightly compressed. There is a small attachment scar at the tip of the base. The pappus is a ring of short membranous teeth extending from the apex. It is ±0.1 mm long and 0.3 mm in diameter.

Surface smooth; dull. At high magnification it appears finely reticulate and glittery. The achene is noticeably hairy with about 6 to 8 distinct rows of bristly white hairs. Dark brown with pale stripes underlying the rows of hairs.

Length 1.9–2.2 mm; thickness 0.8–1.1 mm.
Pastures; dry exposed soil; S 1/3 GP.

222. *Helenium amarum* (Raf.) H. Rock — Bitter sneezeweed

Achene: Outline obtriangular, with a persistent pappus of awned scales from the apex. Cross section variable, often 5-sided. Form obpyramidal with 5 faces and 5 angles. There is a narrow ridge on each angle. The basal attachment point is small, round. There is a short apical collar; the pappus arises just inside the collar. The pappus consists of 5 whitish, translucent, membranous scales, each with a minutely barbed awn. Pappus ±1.6–2.0 mm long.

Surface appears finely roughened at 10× magnification; dull. Many straight hairs, ±0.2–0.6 mm long, arise from the base and lower portion of the angles of the achene. Dark brown; pappus and hairs yellowish white.

Length 1.1–1.3 mm; diameter 0.7–0.8 mm. (Measurements exclude the pappus.)
Waste places; sandy soil; SE 1/8 GP.

223. *Helenium autumnale* L. — Sneezeweed

Achene: Outline triangular with lacerate teeth on the wide end. Cross section round to elliptical. Form conical, slightly compressed, with 5 distinct ribs. The apex is flat. The pappus is a persistent crown of 5 membranous teeth. The teeth are lacerate and awn tipped, ±1 mm long.

Surface conspicuously ribbed; there are long, straight, appressed hairs on the ribs and transparent resinous droplets on the surfaces of the grooves between the ribs. These droplets are visible at 10× magnification. The hairs and droplets are shiny. Orange-brown; the ribs are lighter colored, and the pappus is pale orange.

Length 1.0–1.2 mm; diameter 0.5–0.6 mm. (Measurements exclude the pappus.)

Pastures; moist soil; throughout GP except W 1/4.

224. *Helianthus annuus* L. Common sunflower

Achene: Outline a rounded long triangle. Cross section rhombic or biconvex with narrow wings. Form oblong, compressed, narrowed to a small base and with winglike side margins. Along the center of each face is a prominent broad lengthwise ridge. The apical end is nearly flat. A small, slightly elevated apical collar, ±0.5 mm in diameter, includes a short style base. The pappus is very easily shed. It consists of 2 acute, lacerate white teeth, ±2.5–3.0 mm long, on either side of the apical rim.

Surface initially covered with many fine straight white hairs. These hairs are ±0.5 mm long, oriented lengthwise and uniformly distributed over the surface. However, these hairs are easily worn off; when they are absent, the surface is nearly smooth and has a slight sheen. With magnification, the surface appears striate. There are about 20 fine lengthwise grooves on each broad face of the seed. The spaces between the grooves are rough, and at high magnification appear finely tuberculate. Gray-brown, but whitish when hairs are present; there is a pale smooth area at the base, and the apical rim is whitish.

Length 4.0–4.9 mm; width 2.3–2.6 mm; thickness 1.4–1.9 mm.

Roadsides, cultivated fields, especially corn, sorghum, soybean, and wheat fields; good soils; throughout GP.

225. *Helianthus ciliaris* DC. Texas blueweed

Achene: Outline a rounded oblong, narrowed toward the base. Cross section biconvex to rhombic. Form somewhat oblong, compressed, widest and thickest near the apex and thin near the margins. The basal attachment area is small and oblique. There is a slight apical collar, ±0.5 mm in diameter.

Surface finely striate; shiny. With magnification small rounded tubercles can be seen. Silvery gray or mottled light and dark brown.
Length 2.8–3.2 mm; width 1.8–2.0 mm; thickness 1.1–1.3 mm.
Cultivated fields; damp sandy soil; SW 1/4 GP.

226. *Helianthus petiolaris* Nutt.   Plains sunflower

Achene: Outline obovate. Cross section elliptical to rhombic. Form obovoid, compressed, with a prominent rounded lengthwise ridge on the center of each face. Each face is concave between the ridge and the side margins. The attachment area is oblique at the base. It has a smooth rim with a distinct notch that runs from front to back. There is a very small round apical collar, ±0.7 mm in diameter.

Surface covered with abundant hairs. The hairs are yellowish, soft, and straight, ±1 mm long, and easily worn off. Without the hairs, the surface is smooth and has a slight sheen. At 10× magnification small tubercles in narrow lengthwise rows are visible. Dark gray, with pale mottling near the edges; some achenes are paler and mottled all over.
Length 6.1–7.0 mm; width 2.0–2.9 mm; thickness 1.0–1.5 mm.
Roadsides; sandy soil; throughout GP.

227. *Heliopsis helianthoides* (L.) Sweet   False sunflower,
var. *scabra* (Dun.) Fern.   oxeye

Achene: Outline oblong, narrowed toward the base; often slightly bent to one side. Cross section variable: ray achenes often 3-sided; disk achenes often 4- or 5-sided. Form oblong, with 3 to 5 faces, narrowed to a small base. Basal attachment point very small. There is an apical collar, ±0.5 mm long; the pappus is a short thin lacerate crown on the collar. There is a small beak at the center of the apex.

Surface finely roughened; very dull at 10× magnification. Black. (Reddish when immature.)
Disk achenes: Length 4.2–4.6 mm; diameter 1.2–1.6 mm.
Ray achenes: Length 3.3–4.5 mm: diameter 1.5–2.1 mm.
Waste places; dry soil; E 1/3 GP.

228. *Heterotheca latifolia* Buckl.   Camphor weed

Achene (two distinct forms produced in the same inflorescence):
Ray achene: Outline elliptical, with a blunt base and small-tipped apex. Cross section a narrow wedge. Form unequally 3-faced and 3-angled, elongate, tapering to both the base and the apex. The 2 large elliptical

faces are flat or slightly depressed and meet in a narrow angle. They are joined to a narrow curved face by 2 prominent angles. The circular rimmed attachment scar is at the base. There is no pappus.

Surface finely striate; silky luster. At high magnification a sparse covering of very short brown hairs can be seen. Light yellowish or grayish brown.

Length 2.4–2.7 mm; width 1.1–1.4 mm; thickness 0.4–0.7 mm.
Disk achene: Outline obovate, with a pointed base and a truncate broad end. Cross section very thin. Form obovoid, compressed, thin. There is a very small attachment point at the base. There is an apical rim; a persistent pappus in 2 series arises from the rim. There is a small beak. The pappus consists of a ring of thin white teeth, ±0.5 mm long, and a ring of yellowish minutely barbed capillary hairs, ±0.5 mm long. There is usually a central line or ridge on each face.

Surface with a moderate covering of long straight white hairs, aligned lengthwise. Light brown with darker ridges.

Length 2.1–2.5 mm; width 0.8–1.1 mm; thickness ±0.2 mm.
Pastures, rangeland; sandy disturbed soil; S 1/2 GP.

229. *Iva annua* L. Sump weed, marsh elder

Achene: Outline pear shaped but variable in proportions. Cross section plano-convex. Form like a compressed pear. Usually one face is distinctly convex and the other face flat or slightly concave; the 2 faces meet in a narrow margin. The attachment point is at the small end of the achene. There are 3 lengthwise ridges on each face, though sometimes the center ridge is obscure.

Surface smooth; dull. With magnification it appears finely textured or striate. Medium or dark gray-brown, finely mottled.

Length 2.7–3.8 mm; width 2.1–2.9 mm; thickness 0.8–1.0 mm.
Ditches; moist disturbed soil; SE 1/4 GP.

230. *Iva axillaris* Pursh Poverty weed

Achene: Outline obovate. Cross section elliptical, rounded triangular, or rhombic. Form obovoid, compressed; the dorsal face is convex, and the ventral side has 2 flat faces meeting in a central ridge. Basal attachment area narrow. The apex is nearly flat; there may be a remnant of the style. There is no apical collar or pappus.

Surface scurfy; dull. There may be a few short hairs at the apex. Brown or red-brown; the scurfy layer is pale.

Length 2.6–3.3 mm; width 1.6–2.2 mm; thickness 1.0–1.6 mm.
Rangeland; dry, alkaline, or saline soil; NW 1/4 and W central GP.

231. *Iva xanthifolia* Nutt.　　　　　　　　Marsh elder, sump weed

Achene: Outline obovate. Cross section elliptical. Form clavate, slightly compressed, with a narrow blunt base, a small marginal rim, and a lengthwise ridge from base to apex on each face.

Surface finely textured; dull. At 10× magnification, a pattern of fine shiny tubercles, arranged in lengthwise rows, is visible. There are a few coarse white hairs near the apex. Black, tinged with purple; the basal attachment point is white.

Length 2.5–2.8 mm; width 1.3–1.8 mm; thickness 1.0–1.2 mm.
Ditches; sandy, damp soil; throughout GP.

232. *Kuhnia eupatorioides* L. var. *corymbulosa* T. & G.　　　　　　　　False boneset

Achene: Outline oblong but narrowed toward the base. Cross section variable, generally oblong. Form oblong, compressed, flaring slightly from the base to the apex. There is a smooth knob around the basal attachment point. At the apex there is a small flared apical collar with a persistent pappus of about 20 plumose tan bristles, ±5.5 mm long, and a conspicuous cup-shaped style base.

Surface ribbed, striate; dull. There is a sparse covering of short white hairs. The hairs are more abundant toward the apex. Dark gray to black.

Length 3.2–4.1 mm; width 0.5–0.7 mm; thickness 0.3–0.5 mm.
Pastures, rangeland; sandy soil; throughout GP.

233. *Lactuca canadensis* L.　　　　　　　　Wild lettuce

Achene: Outline elliptical; there may be a long filiform apical beak present. Cross section very thin, often bent. Form ellipsoid, flattened, often somewhat asymmetrical. There is a distinct ridge on each face. The apex of the achene body is cuspidate; a fragile filiform beak extends from the tip. The beak is ±1.5 mm long, flared to a disk at the upper end. From the disk arises a deciduous pappus of many white barbellate bristles, ±6 mm long.

Surface with many fine crosswise wrinkles when seen with low magnification. At 10× these appear as sharply defined, closely spaced irregular ridges. At 25×, each ridge appears to have a row of very short barbs, upwardly directed, on its crest. Mottled brown and black.

Length 3.5–4.4 mm; width 1.8–2.1 mm; thickness ±0.3 mm. (Measurements exclude the beak.)
Margins of woods; partial shade; E 1/2 GP.

234. *Lactuca saligna* L.   **Willow-leaved lettuce**

Achene: Outline elliptical but broadest above the middle; a long fili-form extension of the apical beak may be present, wholly or partly. Cross section a very narrow ellipse. Form ellipsoid, flattened, thin, widest near the apex. There is a very narrow wing on the side margins. There is a smooth rim around the truncate basal attachment point. The filiform portion of the beak is ±2.5 mm long; the readily deciduous pappus consists of many fine white bristles spreading from the end of the beak. The bristles are ±3–5 mm long and appear pointed when seen at high magnification.

Surface with a slight sheen. There are 5 to 8 narrow ribs on each face. The ribs and wings are barbed near the apex. Light brown, with yellow ribs and wings and a white beak.

Length 2.7–3.0 mm; width 0.6–0.8 mm; thickness ±0.1 mm. (Measurements exclude the beak and pappus.)

Gardens, roadsides; disturbed soil; E central GP.

235. *Lactuca serriola* L.   **Prickly lettuce**

Achene: Outline oblanceolate with a small beak. Cross section narrowly elliptical or concave-convex. Form elongate, very thin, narrow at the base and rounded at the apex. The 2 long sides have narrow marginal wings. There are about 5 to 7 lengthwise ribs on each face. At the apex there is a fine white extension, ±3.0 to 3.5 mm long. The pappus of many very fine white bristles, ±3 mm long, arises from the end of the extension. The pappus and all or part of the extension may be broken off.

Surface ribbed; dull. Both the ribs and the marginal wings are minutely upwardly barbed; toward the apex these barbs extend into short straight white hairs. Light olive brown; the ribs and wings are pale yellow.

Length 2.7–3.0 mm; width 0.7–0.9 mm; thickness ±0.1–0.2 mm. (Measurements exclude the extension and pappus.)

Gardens, fallow fields, waste places, wheat fields, pastures; open areas; throughout GP.

236. *Matricaria matricarioides* (Less.) Porter   **Pineapple weed**

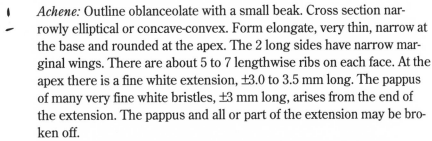

Achene: Outline obovate. Cross section oblong, often 4- or 5-angled. Form elongate, often curved, narrowed toward the base; often with 5 faces and 5 angles; the 2 lateral angles are more pronounced. Basal attachment point small, with a circular rim. Apex obliquely truncate with a slight apical collar. Pappus none.

Surface striate; slight sheen at 10× magnification. Straw-colored, somewhat translucent. Lateral ribs dark orange-brown. Collar opaque.

Length 1.3–1.6 mm; width 0.5–0.7 mm; thickness 0.3–0.4 mm.
Roadsides, lawns; moist open areas; N 1/4 and E 1/4 GP.

237. *Onopordum acanthium* L.   Scotch thistle

Achene: Outline obovate. Cross section variable, generally elliptical to rhombic. Form obovoid, compressed, with slight ridges on the side margins and the center of each face. The basal attachment point is small. The apical area is elliptical, ±1.0 mm long, 0.7 mm wide, with a remnant of the style base. There is a short apical collar, but it is shed with the deciduous pappus. The pappus consists of many barbed bristles, ±6 mm long, straw colored

Surface cross corrugate with distinct wrinkles. The wrinkles are just visible without magnification and conspicuous at 7×; there are about 15 wrinkles on each face. Surface dull without magnification, glaucous at 10×. Wrinkles gray-brown, intervals yellow-brown; entire surface with dark-brown mottles.

Length 4.0–4.9 mm; width 2.2–3.0 mm; thickness 1.5–1.6 mm.
Roadsides, feedlots; rich moist soil; widely scattered in GP.

238. *Silphium integrifolium* Michx.   Whole-leaf rosin-weed

Achene: Outline obovate; at the wide end there are 2 prominent teeth with a large V-shaped notch between them. Cross section very narrow. Form obovoid, compressed, very thin; the whole achene may be curved. There is a small ridge from the base to the apex on each face. A very thin marginal wing, ±1.0 mm wide, extends to form the pointed apical teeth. The notch between the teeth is ±3.5 mm wide and includes the style base.

Surface slightly striate; soft sheen. The wing is stiff and papery, becoming lacerate toward the apex. Fine short hairs may be present on both faces. Gray; the wing is light brown.

Length 1.0–1.1 cm; width 6.1–7.3 mm; thickness ±0.1 mm.
Roadsides; open areas; SE 1/4 GP.

239. *Silphium laciniatum* L.   Compass plant

Achene: Outline a rounded oblong or triangle; at the wide end there are 2 teeth with a rectangular notch between them. Cross

section a narrow curved line, thicker in the center. Form trapezoidal to rounded triangular, compressed, very thin, with the whole achene or the margins curved toward one face. A central lengthwise rib may be present on one or both faces. There is a small pointed tip at the basal attachment point. At the apex are 2 distinct teeth; the rectangular notch between the teeth includes an elliptical depression. The rim of the depression is ragged or has 2 lacerate teeth on each side.

Surface smooth; lustrous. With magnification it appears striate and finely textured. The ventral face, except for the margins, has a dense covering of white strigose hairs. Streaked purple and green; the color of the ventral surface is obscured by the hairs.

Length 1.1–1.2 cm; width 7–9 mm; thickness ±1.1 mm.

Pastures, roadsides; open areas; SE 1/4 and E central GP.

240. *Silphium perfoliatum* L.   Cup plant

Achene: Outline obovate; at the wide end there are 2 small acute teeth with a U-shaped notch between them. Cross section a very narrow ellipse. Form obovoid, compressed, thin, often bent. On each face a rounded ridge extends from the base about 1/2 the length of the achene. A very thin marginal wing, ±0.5 mm wide, extends to form the apical teeth. The notch between the teeth is ±1.5 mm wide and includes the style base.

Surface smooth; soft sheen. It is finely striate and has many small red-brown resinous dots, visible with magnification. The wing is stiff and papery. Gray-brown; the wing is light brown.

Length 8.9–10.3 mm; width 6.2–7.3 mm; thickness ±1.0 mm.

Roadside ditches; moist soil; E 1/4 GP.

241. *Solidago gigantea* Ait.   Late goldenrod

Achene: Outline oblong but slightly narrower toward the base. Cross section 5-sided to round. Form oblong, somewhat 5-sided. There is a small apical collar; the persistent pappus of about 35 minutely barbed tannish white bristles, ±2.8 mm long, arises from the collar.

Surface striate, shiny, with scattered, spreading white hairs, ±0.1 mm long. Yellow-brown.

Length 1.5–1.7 mm; width 0.4 mm; thickness 0.3–0.4 mm.

Pastures; moist soil; throughout GP.

242. *Solidago rigida* L.   Stiff goldenrod

Achene: Outline oblong but narrowed toward the base. Cross section oblong or 5-sided. Form oblong, compressed. There is a small thin flaring apical collar; the persistent pappus of minutely barbed white hairs, ±4 mm long, spreads from the collar.

Surface with closely spaced smooth lengthwise ridges; there are about 8 ridges on each of the 2 broad faces, and the interspaces are very narrow. Ridges glossy. The ridges are pale tan, the interspaces brown.

Length 2.4–3.0 mm; width 0.5–0.9 mm; thickness 0.3–0.5 mm.
Pastures; dry sandy soil; throughout GP except SW 1/8.

243. *Sonchus arvensis* L.   Field sow thistle, perennial sow thistle

Achene: Outline elliptical but narrower toward the base; both base and apex truncate. Cross section elliptical. Form ellipsoid, compressed; side margins thin but not winged. There is a short, flaring apical collar, ±0.4 mm in diameter. The deciduous pappus of many fine white hairs arises from the collar. The pappus hairs are ±8–10 mm long; they appear minutely barbed at 10× magnification.

Surface conspicuously ridged with ±12–15 lengthwise ribs; dull. The ribs are rough with coarse teeth; the grooves between the ribs are smooth. Orange-brown.

Length 2.5–2.9 mm; width 0.8–1.1 mm; thickness 0.5–0.6 mm.
Cultivated fields; good soils; N 1/4 and E central GP.

244. *Sonchus asper* (L.) Hill   Prickly sow thistle, spiny sow thistle

Achene: Outline somewhat elliptical, gradually tapered to the base and abruptly narrowed at the apex. Cross section a very thin ellipse. Form ellipsoid, compressed; the margins are winged, and there are 3 fine lengthwise ribs on each face. There is a small white apical collar, ±0.2 mm in diameter. The pappus of many fine soft white hairs, ±7 mm long, spreads from the collar.

Surface smooth; dull. When seen at 10× magnification it appears very finely roughened and has a slight sheen. At high magnification minute barbs on the marginal wings and a few barbs on the ribs are visible. Light to medium red-brown, with paler wings and ribs.

Length 2.7–3.0 mm; width 1.3–1.5 mm; thickness ±0.3 mm.
Cultivated fields, gardens; disturbed soil; throughout GP.

245. *Taraxacum laevigatum* (Willd.) DC. **Red-seeded dandelion**

Achene: Outline elliptical but narrowed toward the base. Cross section biconvex or concave-convex, thin and ridged. Form ellipsoid, compressed, sometimes curved. There are about 10 lengthwise ridges on the achene, interspersed with about 5 grooves, all irregularly spaced. The base is truncate with a small dark sunken attachment scar. A filiform extension, ±7–8 mm long, projects from the apex but is easily broken. The pappus consists of many hairs spreading from the end of the extension. These hairs are ±5 mm long, white, and finely barbed.

Surface finely textured; dull. Each of the ridges has about 5 thin upwardly directed teeth. These teeth are larger and more flared toward the apex. Orange-brown.

Length 3.8–4.0 mm; width 0.7–0.9 mm; thickness ±0.4 mm. (Measurements exclude the extension and pappus.)

Lawns; disturbed soil; throughout GP.

246. *Taraxacum officinale* Weber **Common dandelion**

Achene: Outline elliptical. Cross section a thin oblong. Form oblong or ellipsoid, compressed, with about 10 lengthwise ridges. A filiform extension, ±6.5–7.0 mm long, projects from the apex but is easily broken. The pappus consists of many fine hairs spreading from the tip of the extension. These hairs are ±5 mm long, white, and finely barbed.

Surface rough; dull. When seen at high magnification it appears finely striate. The ridges are slightly toothed near the basal end, with the teeth becoming more prominent toward the apex. In the apical 1/3 the teeth are flaring and upwardly directed. Dull olive.

Length 3.3–4.0 mm; width 0.6–1.0 mm; thickness ±0.4 mm. (Measurements exclude the extension and pappus.)

Lawns; disturbed soil; throughout GP.

247. *Tragopogon dubius* Scop. **Goat's beard**

Achene: Outline long elliptical, often curved, with an oblique base and a long apical beak with long, spreading hairs. Cross section 5-angled. Form ellipsoid, very long, somewhat 5-sided, variously bent or curved, with an oblique base and an extended beak. The base is hollow and has a small remnant of a stalk. There are 10 lengthwise ribs; 5 of these form distinct angles that alternate with 5 less-defined ribs. From the beak arise many firm bristles, ±2.3–2.7 cm long. These bristles are feathery with short soft hairs. The beak is usually persistent on the achene, but the bristles are readily shed.

Surface rough; dull. All 10 ribs are covered by thin, flaring, upwardly directed teeth. At 20× magnification these teeth appear to be covered with tiny scales that are oriented lengthwise. Buff to dull brown, with the beak, bristles, and teeth paler.

Length (including beak) 2.9–3.4 cm; length (toothed portion only) 1.2–1.4 cm; thickness 1.5–1.6 mm.

Roadsides, grain fields; disturbed soil; throughout GP.

248. *Vernonia baldwinii* Torr.   Western ironweed

Achene: Outline oblong, narrowed toward the base. Cross section elliptic or 5-sided. Form terete or slightly compressed, narrowed slightly to the blunt base. There is a smooth rim around the basal attachment point. The apex is flat, and the pappus is persistent on a very short apical collar. The pappus consists of 2 series of shiny purple barbed bristles. The inner series is ±6 mm long; the outer series is ±1 mm long. There are about 5 narrow ribs on the body of the achene.

Surface dull. There are very fine hairs on the ribs, visible with high magnification. In the intervals between the ribs are many small clear resinous droplets. Yellow-brown; there are some dots or streaks of purple, mostly on the intervals.

Length 2.4–2.8 mm; width 0.8–1.1 mm; thickness 0.6–0.9 mm. (Measurements exclude the pappus.)

Pastures, rangeland; dry soil; S 1/2 GP except W 1/4.

249. *Vernonia fasciculata* Michx.   Ironweed

Achene: Outline oblong, narrowed toward the base. Cross section round or elliptical, distinctly toothed. Form terete or slightly compressed, narrowed slightly to the blunt base. There is a circular rim at the basal attachment point. The apex is flat, and the pappus is persistent on a very short apical collar. The pappus of shiny purple barbed bristles is in 2 series. The inner series is ±6 mm long; the outer series is ±1–2 mm long. There are about 10 prominent narrow ribs on the body of the achene.

Surface dull. There are some very fine hairs on the ribs, visible with high magnification. In the intervals between the ribs are many small yellow resinous droplets. Yellow-brown; there are small flecks of purple in the intervals.

Length 3.0–3.6 mm; width 0.8–1.0 mm; thickness 0.6–1.0 mm. (Measurements exclude the pappus.)

Pastures; damp disturbed soil; E central GP.

250. *Xanthium spinosum* L. Spiny cocklebur

Bur (tough spiny structure derived from the involucre; has 2 chambers, each enclosing 1 achene): Outline elliptical, spiny. Cross section elliptical. Form ellipsoid, slightly compressed, with a blunt base. At the apex are 2 acute beaks, ±0.5 mm long; typically, just below one of the beaks is a stout straight spine, 1–3 mm long. There are many spreading spines, ±1.5–2.0 mm long; these are straight except for a distinct hook-shaped curve at the tip.

Surface uneven, spiny. It has fine lengthwise ridges but is mostly covered with many tangled short white hairs and yellow resinous droplets. Light yellow-brown.

Length 1.0–1.2 cm; width ±7.0 mm; thickness ±6.0 mm. (Measurements include the spines and beaks.)

The achenes are elliptical in outline and plano-convex in cross section. Dark gray, striate. Length ±9 mm; width ±2 mm; thickness ±1 mm.

Irrigated fields, waste places; damp rich soil; S 1/3 GP.

251. *Xanthium strumarium* L. Cocklebur

Bur (tough spiny structure derived from the involucre; has 2 chambers, each enclosing 1 achene): Outline long ovate, with many hooked spines and with 2 hooked teeth spreading from the apex. Cross section round. Form long-ovoid, tapered to a small blunt attachment point at the base, very spiny. There are 2 beaks, each ±5 mm long and curved inward to a hooked tip; the beaks form a V at the apex.

Surface dull, covered with many curly hairs, spiny. The bases of both the spines and the beaks are also hairy. Light brown.

Length 2.2–2.8 cm; diameter 1.4–1.7 cm. (Measurements include the spines and beaks.)

The achenes are compressed ellipsoid in form, broadest below the middle, with a small basal attachment point. Cross section plano-convex. The seed itself is tan, but it is enclosed in a striate papery pericarp that is silvery-black. The pericarp has lengthwise ridges and is tapered to a fine tip at the apex. Length ±10 mm; width ±3 mm; thickness 1.2 mm.

Cultivated fields, especially corn, sorghum, and soybean fields, rangeland, waste places; moist or drying soil; throughout GP.

Poaceae
Grass Family

252. *Aegilops cylindrica* Host. **Jointed goat grass**

Spikelet (usually found as a spikelet of 2 or 3 florets firmly attached to a segment of the rachis): The spikelet and associated rachis form a cylindrical unit. The rachis has ±9 slightly barbed nerves; it flares and becomes thicker toward the top. The spikelet fits closely into the rachis. Two nearly equal glumes and the awns of 2 lemmas are visible. The glumes are thick and firm, with ±7 slightly barbed nerves. Each glume has a bifid apex with the outer tooth expanded into a barbed awn. The lemma is membranous, firm; there is a short awn from the back of the lacerate, bifid apex. The palea is membranous with ciliate margins. It is adherent to the caryopsis. The length of the awns varies greatly with the position on the rachis. Awns of the glumes range from ±4 mm long (lower spikelets) to ±10 mm (upper spikelets), with that of the uppermost spikelet much longer, ±6 cm. The awns of the lemmas range from ±1.5 mm to 7 mm long. Straw colored or greenish.

Length (entire unit) 9.5–10.5 mm; diameter ±3 mm.

Wheat and other grain fields, roadsides, waste places; open areas; S 1/2 GP.

253. *Agropyron repens* (L.) Beauv. **Quackgrass**

Floret (may be found as a floret or as a spikelet of several florets): Outline elliptical, broadest below the middle. Cross section C-shaped. Form ellipsoid, convex on the dorsal side and concave on the ventral side. The caryopsis has a C-shaped cross section. The lemma (convex side) is ±9–10 mm long, 1.0–1.7 mm wide, with hyaline margins and a scabrous awn from the apex. The palea, ±7.0–7.5 mm long, has infolded margins; the folds are minutely barbed. The basal attachment area is oblique, 3-sided. The rachilla may be present; it is hairy and ±1.0 mm long.

Both the lemma and palea are thin and firm, with obscure nerves. At high magnification the lemma appears very finely tuberculate. Straw colored; the caryopsis is brown.

Length ±10.0 mm; width ±1.7 mm.

Lawns, cultivated fields; moist soil; N 2/3 GP.

254. *Agropyron smithii* Rydb. **Western wheat grass**

Floret: Outline elliptical, broadest below the middle. Cross section C-shaped. Form ellipsoid; the dorsal (lemma) side convex, the ventral (palea) side concave. Lemma firm, with hyaline margins, ±7 nerves, some of them faint; with an awn, ±1–2 mm long, from the apex. Palea shorter than lemma, its center concave, the sides strongly reflexed. The folds on the sides of the palea are ridged and minutely barbed. The rachilla is persistent, 1.5–2.0 mm long. The basal area is oblique, triangular.

Surface dull. At 20× magnification the palea and inner surface of the lemma are short-hairy. Straw colored.

Length (excluding awn) 6.9–10.5 mm; width 1.3–1.7 mm.

Spikelet: Several florets subtended by subequal glumes. The glumes are lanceolate, short awned, ±10 mm long. The caryopsis is oblong, with C-shaped cross section, apex hairy; ±4.5 mm long, 1.4 mm wide.

Roadsides; dry soil; throughout GP.

255. *Bromus inermis* Leyss. subsp. *inermis*   Smooth brome

Floret: Outline elliptical, widest above the middle. Cross section a slightly curved line. Form ellipsoid, thin, flat. The lemma is 3-nerved and has very thin marginal wings that are inrolled in the lower half. The palea is 3-nerved, slightly shorter than the lemma, and adherent to the caryopsis. The basal attachment area is oblique, with the circular scar visible on the palea side. The rachilla is usually present and is about 1/3 the length of the floret.

The lemma and palea appear smooth and membranous. The rachilla and the nerves of the lemma and palea are minutely scabrous, with short hairs that are visible at 10× magnification. Light yellow-brown, with the dark brown caryopsis evident through the palea. The lemma margins are transparent; the rachilla and attachment area are pale.

Length 9.5–10.6 mm; width 1.9–2.7 mm; thickness 0.4–0.5 mm.

Roadsides; grain fields; open areas; throughout GP.

256. *Bromus japonicus* Thunb. ex Murr.   Japanese brome

Floret (may be found as a floret or as a spikelet of 6–10 florets): Outline a narrow ellipse with a long awn from the apex. Cross section C-shaped. Form long-ellipsoid, with 1 concave face. The lemma, on the convex face, is rhombic, with a bifid apex. It is 7-nerved and has transparent inrolled margins. A minutely barbed awn, ±1.0–1.2 cm long, arises from the back of the lemma, about 1/4 the distance below the apex. The palea is thin and adherent to the caryopsis. The base of the floret is oblique, with a round attachment scar. The rachilla may be present in the groove on the palea side.

The lemma and palea are membranous and smooth and have a slight luster. There are scattered very short hairs on the lemma; the palea has scattered marginal hairs, ±0.5 mm long. Straw colored.

Length 7.9–9.0 mm; width 1.0–1.6 mm. (Measurements exclude the awn.)

Grain fields, especially winter wheat, waste places; moist or dry soil; throughout GP.

257. *Bromus secalinus* L.   Cheat

Floret: Outline elliptical. Cross section U-shaped. Form ellipsoid, folded lengthwise, with a deep groove on the ventral side. The lemma is rounded and slightly keeled. It has a straight or reflexed scabrous awn, ±5–7 mm long, that arises just below the apex. The palea has ciliate margins; it is thin and adherent to the caryopsis. A smooth white rim at the base is visible from the dorsal side. On the ventral side there is a circular attachment scar at the base. The rachilla is conspicuous, ±1.5 mm long; it lies above the groove and is curved outward.

The lemma (outer surface of the floret) is thin and membranous, with translucent margins. Tan or greenish; the caryopsis is brown.

Length 7.5–7.9 mm; width (side to side) 0.9–1.7 mm; thickness (dorsal-ventral) 1.7–2.1 mm.

Winter wheat and alfalfa fields; good soils; SE 1/4 GP.

258. *Bromus tectorum* L.   Downy brome

Floret (may also be found as a spikelet of ±4–7 florets): Outline narrowly elliptical, widest just above the middle, and with a long awn from the apex. Cross section C-shaped, form long-ellipsoid, thin. The floret appears to be rolled around the lengthwise axis and then reflexed. The lemma is elliptical with a bifid apex and thin transparent margins. It has 7 nerves. A minutely barbed stiff awn arises from the back of the lemma, ±2 mm below the apex. The awn may be straight or reflexed. The palea is elliptical and 3-nerved. There is a thin smooth area at the base, with a round attachment scar. The rachilla, ±2 mm long, may be present.

The lemma and palea are membranous and dull. The lemma has straight white hairs, with longer hairs near the margins. The palea has scattered long hairs on its margins. Tan, with a red tinge.

Length 10.5–12.1 mm; width 1.1–1.5 mm. (Measurements exclude the awn.)

Winter wheat and alfalfa fields, waste areas, fence rows; moist soil; throughout GP.

259. *Chloris verticillata Nutt.* Windmill grass

Caryopsis: Outline elliptical. Cross section rounded triangular. Form long-ellipsoid, 3-sided, with a curved ventral face and 2 flat faces on the dorsal side. The scutellum is a depressed area at the base of the dorsal side. The scutellum extends 1/3 to 2/3 the length of the seed. It has a central lengthwise ridge and a very small rim.

The surface is finely striate and appears somewhat glossy when seen with magnification. Translucent yellow; the basal end of the curved face is brown.

Length 1.4–1.6 mm; thickness 0.5–0.6 mm.

Cultivated fields, gardens, roadsides; sandy or gravelly dry soil; S 1/2 GP.

260. *Cynodon dactylon* (L.) Pers. Bermuda grass

Floret: Outline somewhat elliptical; 1 long edge is nearly straight, the other distinctly curved. Cross section a narrow triangle. Form ellipsoid, compressed, with 2 broad faces and 1 very narrow face. The lemma encloses the broad faces and is keeled along the curved long edge. The palea clasps the narrow face. The rachilla lies along the palea and is about 3/4 the length of the floret.

The lemma and palea appear slightly striate and have a slight sheen. At high magnification a striate-reticulate pattern is visible. The keel of the lemma is hairy. Purplish brown.

Length 1.9–2.2 mm; width 0.8–0.9 mm; thickness 0.4–0.5 mm.

Pastures, waste places, lawns, gardens; moist areas; S 1/2 GP.

261. *Digitaria ciliaris* (Retz.) Koel. Southern crabgrass

Spikelet of 2 florets, the lower floret sterile: Outline long-elliptical with a blunt base and narrow apex, widest below the middle. Cross section plano-convex. Form long-ellipsoid, broadest below the mlddle, with 1 flat face. The first glume is ±0.4–0.5 mm long. The second glume is 3-nerved, narrow, ±2.2 mm long; it has conspicuous marginal hairs. The sterile lemma is as long as the spikelet; it has 3 nerves and ciliate margins. The glumes and sterile lemma are thin and membranous. The fertile lemma is firm and has an acuminate tip and inrolled membranous margins that clasp the firm palea. Only a narrow triangular area of the palea is visible.

121

The glumes and sterile lemma are dull. Greenish, tinged with purple. The lemma and palea are smooth and slightly shiny. When seen with magnification the surface of the lemma appears finely roughened with lengthwise rows of tiny pits. Both the lemma and the palea are straw-colored or greenish.

Length 3.0–3.6 mm; width 0.9–1.0 mm; thickness 0.4–0.6 mm.

Lawns, cultivated fields, especially corn, soybean, and sorghum fields; moist soil; S 1/2 GP.

262. *Digitaria ischaemum* (Schreb. ex Schweigg.) Schreb. ex Muhl.   Smooth crabgrass

Spikelet of 2 florets, the lower floret sterile: Outline elliptical with a blunt base and pointed apex. Cross section plano-convex. Form ellipsoid with 1 flat face. The first glume is lacking. The second glume and the sterile lemma are both membranous, greenish, lightly hairy, and as long as the floret. The glume is 3-nerved and the sterile lemma 5-nerved. The fertile lemma and the palea are thin but firm and shiny. The very thin inrolled margins of the lemma clasp the palea, permitting only a small strip of the palea to show.

The surface of both lemma and palea appears striate when seen at low magnification but distinctly scalariform at high magnification. The lemma and palea are black. The lemma margins are whitish.

Length 1.9–2.0 mm; width 0.9–1.0 mm; thickness 0.4–0.5 mm.

Lawns, cultivated fields, especially corn, soybean, and sorghum fields; moist soil; E 1/2 GP.

263. *Digitaria sanguinalis* (L.) Scop.   Hairy crabgrass

Spikelet of 2 florets, the lower floret sterile: Outline long elliptical with a lanceolate tip and blunt base. Plano-convex in cross section. Form long ellipsoid, flat on 1 face and with a long tapered apex. The first glume is very small, ±0.4 mm long, with an acute tip. The second glume is more than 1/2 as long as the spikelet and has 3 nerves, hairy margins, and a lanceolate tip. The sterile lemma is as long as the spikelet, with 3 prominent nerves and ciliate margins. The glumes and sterile lemma are membranous. The fertile lemma and palea are firm and clasp the caryopsis securely.

The glumes and sterile lemma are tan and dull. The lemma and palea are nearly smooth, with fine striations, dull olive in color, and with a slight sheen.

Length 2.7–3.0 mm; width 0.8–0.9 mm; thickness 0.4–0.5 mm. (Measurements of entire spikelet.)

Lawns, cultivated fields, especially corn, soybean, and sorghum fields; moist soil; S 2/3 GP.

264. *Echinochloa crus-galli* (L.) Beauv.   Barnyard grass

The unit of dispersal is a floret or a spikelet of 2 florets; the lower floret is sterile.

Floret: Outline elliptical with a long tapered apex. Cross section plano-convex. Form ellipsoid with 1 long tip. One face is flat; the other face is strongly convex or has a hump in the center but is thin toward the apex. The lemma margins clasp the palea, enclosing the caryopsis.

The lemma and palea are smooth and very shiny. They are tan with faint striate markings. The lemma has 3 to 5 faint, pale lengthwise stripes.

Length 3.5–4.2 mm; width 1.6–2.0 mm; thickness 1.1–1.4 mm.

Spikelet: The first glume (on the flat side) is about 1/2 as long as the spikelet, 3-nerved, with an acute tip. The second glume is as long as the spikelet, 3-nerved, with a tapered tip. The sterile lemma is as long as the spikelet and clasps the fertile floret. It is 3-nerved, with the central nerve extended into a very long awn. The glumes and sterile lemma are membranous. They are covered with short hairs and have spiny nerves.

Cultivated fields, especially corn, alfalfa, soybean, and sorghum fields; moist disturbed soil; throughout GP except NW 1/4.

265. *Eleusine indica* (L.) Gaertn.   Goosegrass

Caryopsis: Outline elliptical. Cross section heart shaped. Form elongate, somewhat 3-sided. There are 2 flat faces that join at about 45° in a rounded angle. The third face (the ventral side) has a distinct V-shaped lengthwise furrow. The apex is rounded. Near the base on the furrowed side is a flattened area with a round, slightly elevated attachment scar, ±0.1 mm in diameter. At the base on the angled side the scutellum is an indented area oblique to the angle. The scutellum and the flattened area together form a blunt tip at the base of the caryopsis.

The surface is covered with 2 sets of arched concentric ridges. These arches begin at the scutellum, rise toward the apex on each side, and descend to the furrow. There are about 10 arches on each face. The ridges are narrow and wavy. At high magnification a reticulate pattern is visible between the ridges, and the surface appears shiny. Dark orange-brown. At high magnification the ridges are dark and the interspaces are light or medium orange.

Length 1.0–1.3 mm; width 0.4–0.5 mm; thickness 0.5–0.6 mm. (Width measured across both flat faces.)

Roadsides, gardens; compacted soil; SE 1/4 GP.

266. *Elymus canadensis* L.   Canada wild rye

Floret: Outline elliptical with a very long straight awn from the apex. Cross section ellipsoid, with a rectangular notch in one long edge. Form ellipsoid, elongate, flattened. The lemma is 3-nerved and has a straight scabrous awn, ±3 mm long, extending from the apex. The margins of the lemma curve over the palea. The palea has a rounded apex and recurved side margins that form a groove. The base is oblique, with the attachment point visible on the palea side. The rachilla and sometimes a sterile floret may be present in the groove.

The lemma and palea are membranous and dull. Tan. The lemma has fine short white hairs; the palea is glabrous.

Length (not including awn) 6.8–7.5 mm; width 1.1–1.6 mm; thickness 0.8–1.0 mm.

Roadsides, cultivated fields; sandy or gravelly soil; throughout GP.

267. *Eragrostis cilianensis* (All.) E. Mosher   Stinkgrass

Caryopsis: Outline ovate. Cross section round or slightly compressed. Form ovoid with a tiny tip at the base; slightly compressed sideways. The scutellum is a depressed area on the dorsal face, extending from the base 1/3 or more the length of the caryopsis. There is a distinct narrow ridge in the center of the scutellum.

Surface smooth, dull. At high magnification a low relief reticulate pattern is visible. Orange-brown. The attachment point is a dark spot near the base.

Length 0.5–0.6 mm; width 0.4–0.5 mm; thickness ±0.4 mm.

Waste places, corn and sorghum fields, roadsides; dry gravelly soil; throughout GP.

268. *Festuca octoflora* Walt.   Six-weeks fescue

Floret (disarticulation above the glumes, so none present): Outline elliptical with a truncate base and tapered, awned apex. Cross section C-shaped. Form narrowly ellipsoid, rolled, with the dorsal face convex and the ventral face deeply grooved. The lemma (dorsal face) is in-rolled to meet the edges of the concave palea. An awn, ±0.5–0.7 mm long, extends from the apex of the lemma. The rachilla is usually present on the palea side; it is ±0.5 mm long and has a broad flat apex. The basal attachment area is smooth and elliptical.

The lemma is minutely scabrous. Yellow-brown, translucent, with the caryopsis visible as an orange area.

Length (including awn) 2.6–3.9 mm; width 0.5–0.6 mm; thickness 0.4–0.5 mm.

Roadsides; dry sterile soil; throughout GP.

269. *Hordeum pusillum* Nutt.   Little barley

Spikelet (usually found as a fertile spikelet between 2 sterile spikelets on a segment of the rachis): Outline long-elliptical with a long awn from the apex. Cross section plano-convex. Form long-ellipsoid, with a blunt base and long tapered apex; flat on 1 face and widest just below the middle. Both glumes and the lemma of the fertile floret are awned, the awns as long as 7 mm. The lemma clasps the palea. The palea has a lengthwise groove; a rachilla segment is in the groove. The rachilla segment is about 3/4 the length of the floret. The sterile spikelets are pedicellate and narrowly elliptical. The first glumes are lanceolate and awned, the second glumes awnlike. The lemmas and paleas are narrow.

Surface of the lemma striate; it has fine, short bristles. The palea is smooth. Both lemma and palea are papery and dull. Buff colored.

Length 5.9–7.0 mm; width 1.6–1.8 mm; thickness 1.0–1.2 mm. (Measurements of the fertile spikelet, excluding the awn.)

Rangeland, roadsides, pastures; dry or alkaline soil; S 2/3 GP.

270. *Panicum capillare* L.   Common witchgrass

Floret: Outline elliptical, with an acute apex and blunt base. Cross section rounded elliptical. Form ellipsoid, compressed front to back, and with one face somewhat flattened. The lemma, on the more convex face, has 5 nerves. The lemma margins clasp the 2-nerved palea. At the base is a small crosswise elongate attachment scar.

Both lemma and palea are smooth and very shiny. Both are dark gray with pale yellow nerves, margins, and base. When seen at high magnification, the gray areas appear finely streaked.

Length 1.3 mm; width 0.8 mm; thickness 0.6–0.7 mm. (Measurements of entire floret.)

Fallow fields, corn, sorghum, and soybean fields, waste places; sandy or rocky exposed soil; throughout GP.

271. *Panicum dichotomiflorum* Michx.   **Fall panicum**

The unit of dispersal is a floret or a spikelet consisting of a lower sterile floret and an upper fertile floret.

Spikelet: Elliptical with a narrow apex. The glumes and sterile lemma are loose and papery in texture. They are buff to green in color or sometimes purplish. The first glume is blunt, 1/4 to 1/3 as long as the spikelet. The second glume has 5 to 7 distinct nerves. The sterile lemma has about 5 nerves. Length of the spikelet is ±2.6 mm.

Floret: Outline elliptical, with a blunt base and acute tip. Cross section elliptical. Form ellipsoid, somewhat compressed, with 1 face convex and the other nearly flat. The lemma, on the convex face, clasps the palea, enclosing the grain. The lemma has 5 distinct nerves and 2 faint nerves that all converge toward the apex. The 3 nerves on each side of the central nerve originate from a lateral point near the base.

The lemma and palea are smooth and shiny. With magnification the surface appears striate. Light to dark olive; the nerves, apex, and base are paler.

Length 2.0–2.1 mm; width 0.8–0.9 mm; thickness 0.5–0.6 mm. (Measurements of the floret.)

Corn, sorghum, and soybean fields, roadsides; moist soil; SE 1/4 and E central GP.

272. *Panicum miliaceum* L.   **Broom-corn millet, wild proso millet**

Floret: Outline elliptical. Cross section elliptical. Form short-ellipsoid, compressed, with a blunt base and acute apex. It is thickest above the middle, appearing obovate in edge view. The V-shaped basal attachment scar is visible from the palea side. The lemma and palea are firm; they completely enclose the caryopsis. The lemma has 7 faint nerves: 1 central, 1 near each margin, and a pair on each side of the center, all converging at the apex. The palea has 4 nerves.

Both lemma and palea are very smooth and shiny. When viewed at high magnification, a very fine low-relief scalariform pattern is visible. Orange; the margins of the lemma are pale.

Length 2.9–3.2 mm; width 2.0–2.2 mm; thickness 1.5–1.7 mm. *Cultivated fields; disturbed or irrigated soil; N 2/3 GP.*

273. *Panicum virgatum* L. Switchgrass

Floret: Outline long-ovate. Cross section plano-convex. Form long ovoid with one flat face. The lemma, on the convex side, has margins that clasp the palea. The lemma margins are shaped so that the exposed area of the palea has an hourglass shape. The base is obliquely truncate, making the round attachment scar visible on the flat face. From the convex side, the base appears as a small smooth tip.

The lemma and palea are smooth and shiny. Faint lengthwise striations are visible with magnification. Both the lemma and palea are pale straw colored; 5 pale nerves may be visible on the lemma.

Length 2.8–3.0 mm; width 0.9–1.1 mm; thickness 0.6–0.8 mm.

Pastures; moist soil; throughout GP.

274. *Setaria faberi* Herrm. Giant foxtail

The unit of dispersal is a floret or a spikelet composed of a lower sterile floret (lacking glumes) and an upper fertile floret.

Spikelet: Ellipsoid overall. The first glume is ±1 mm long, 3-nerved. The second glume (convex face) is ±2 mm long, 5-nerved. The sterile lemma is the full length of the floret, 5-nerved, with inrolled margins. The glumes and sterile lemma are membranous.

Floret: Outline elliptical, with acute ends. Cross section plano-convex. Form ellipsoid, with 1 nearly flat face and 1 strongly convex face. In edge view, the floret appears thickest at or just below the middle. The lemma, on the convex face, clasps the palea, completely enclosing the caryopsis. The attachment scar is at the base on the flat side.

Both the lemma and the palea are hard and shiny. With magnification, small paplllae arranged in lengthwise rows are visible. The lemma also has fine cross-corrugations. On the palea there is a narrow smooth lengthwise strip on either edge, next to the lemma margins. Usually gray-brown but may be light brown or pale green. There may be 2 pale lengthwise stripes on the lemma.

Length 2.5–2.8 mm; width 1.4–1.5 mm; thickness 1.0–1.2 mm.

Waste places, cultivated fields; moist or disturbed soil; central and E central GP.

275. *Setaria glauca* (L.) Beauv. Yellow foxtail

The unit of dispersal is a floret or a spikelet composed of a lower sterile floret (lacking glumes) and an upper fertile floret.

Spikelet: Ellipsoid. The first glume is ±1 mm long, 3-nerved. The second glume (convex face) is ±1.5 mm long with a blunt tip, 3-nerved. The sterile lemma is the full length of the floret, 5-nerved. The glumes and sterile lemma are thin and papery, greenish.

Floret: Outline elliptical with acute ends. Cross section plano-convex with the convex side very strongly so. Form short ellipsoid with one flat face. The lemma, on the convex face, clasps the palea, enclosing the caryopsis. The attachment point is at the base.

Both the lemma and the palea are firm and dull. With magnification they appear somewhat shiny. Both are finely papillate. The lemma is cross-corrugate with rough ridges that are distinct when seen with magnification. The palea is finely striate and slightly cross-corrugate. Dark gray-brown with paler ridges, base, and lemma margins.

Length 2.6–2.8 mm; width 1.5–1.6 mm; thickness 1.0–1.2 mm.

Corn, sorghum, and alfalfa fields, roadsides, lawns; disturbed soil; E 2/3 GP.

276. *Setaria verticillata* (L.) Beauv.   **Bristly foxtail**

The unit of dispersal is a floret or a spikelet composed of a lower sterile floret (lacking glumes) and an upper fertile floret.

Spikelet: Outline elliptical. Cross section somewhat plano-convex. Form ellipsoid. The first glume is ±1.0 mm long, acute, 1-nerved. The second glume (convex face) is as long as the fertile floret, 7-nerved. The sterile lemma is as long as the fertile floret; the sterile palea is very thin, about half as long the sterile lemma. The glumes and sterile floret are all membranous.

Floret: Outline elliptical, broadest below the middle. Cross section plano-convex. Form ellipsoid, with 1 flat face. The lemma, on the convex face, clasps the palea, completely enclosing the caryopsis and leaving an oblong area of the palea exposed. The attachment point is at the base.

Surface of both lemma and palea striate, with a slight sheen at 10× magnification. At 20× magnification, the surfaces are tuberculate in a lengthwise pattern, and the lemma is slightly cross-corrugate. Straw colored.

Length 2.3–2.5 mm; width 1.0–1.2 mm; thickness 0.6–0.8 mm.

Cultivated fields, waste areas; disturbed soil; NE 1/4 and central GP.

277. *Setaria viridis* (L.) Beauv.   Green foxtail

The unit of dispersal is a floret or a spikelet composed of a lower sterile floret (lacking glumes) and an upper fertile floret.

Spikelet: The first glume is ±0.7 mm long, 3-nerved. The second glume (convex face) is nearly the length of the floret, 5-nerved. The sterile lemma is the full length of the floret, 5-nerved. The glumes and sterile lemma are thin and papery.

Floret: Outline elliptical with acute ends. Cross section plano-convex. Form ellipsoid with 1 flat face. There is a slightly elevated round area near the base on the convex face. The lemma (on the convex face) clasps the palea, enclosing the caryopsis. The attachment point is at the base.

Both the lemma and the palea are firm and appear shiny when seen with magnification. There are 3 faint longitudinal ridges on the lemma. Both the lemma and the palea have a fine striate pattern of papillae. When seen at high magnification, the lemma and palea appear slightly cross-corrugate. The palea has a smooth, shiny lengthwise strip on either edge, next to the lemma margins. Usually tan with paler ridges but may vary from pale green to brown.

Length 1.8–1.9 mm; width 1.0 mm; thickness 0.6–0.7 mm.

Corn, sorghum, and alfalfa fields, roadsides; disturbed soil; throughout GP.

278. *Sorghum bicolor* subsp. *drummondii*   Shattercane
(Steud.) de Wet

Spikelet (usually found as a single fertile spikelet with 2 pedicellate sterile spikelets attached): Outline short elliptical, the apex acute. Cross section rounded triangular. Form ellipsoid. The glumes completely enclose the floret. The dorsal face is somewhat flattened; the ventral face bulges outward. The side margins of the first glume (dorsal face) are folded around the second glume. The margins of the second glume are inrolled. The lemma and palea are membranous, very thin. There are 2 pedicels appressed to the ventral face. Each pedicel may bear a lanceolate sterile spikelet. These are nerved, membranous, ±3.5 mm long, and slightly exceed the fertile spikelet. The caryopsis is ellipsoid, dull, brown, and ±3 mm long.

Surface of the glumes very hard, smooth, and shiny; ventral face slightly ribbed. There are stiff fine yellowish or reddish hairs that are more abundant near the base, apex, and side margins. The pedicels have abundant silky hairs. Dark red-brown to black, with scattered white granules.

Length 4.3–5.5 mm; width 2.8–3.2 mm; thickness 2.4–2.8 mm.
Corn and soybean fields, waste places; disturbed soil; S 1/2 GP.

279. *Sorghum halepense* (L.) Pers.   Johnson-grass

Spikelet (usually found as a single fertile spikelet with 2 pedicellate sterile spikelets attached): Outline elliptical, the apex acuminate; broadest at or just below the middle. Cross section rounded triangular. Form ellipsoid. The glumes completely enclose the floret. The first glume (dorsal face) is flat; the second glume has a central lengthwise hump that extends into a ridge near the apex. The side margins of the first glume are folded around the second glume. The margins of the second glume are inrolled. The lemma and palea are membranous and very thin. There may be a bent awn from the apex of the lemma, but it is readily deciduous. One or 2 pedicels are appressed to the ventral face; each may bear a sterile spikelet that is lanceolate, membranous, ±4.5 mm long, exceeding the fertile spikelet. The caryopsis is obovate, dull, brown, and ±2 mm long.

Surface of the glumes firm, smooth, and shiny. A fine scalariform pattern is visible at high magnification. There are silky hairs on the pedicels and a few on the glumes of the fertile spikelet. Straw colored or purplish or with purple streaks or blotches.

Length 4.5–4.8 mm; width 1.5–2.0 mm; thickness 1.1–1.4 mm.
Corn, sorghum, and winter wheat fields, roadsides, ditches; moist soil; S 1/2 GP.

280. *Sporobolus neglectus* Nash   Poverty grass, dropseed

Caryopsis: Outline generally elliptical, with a small tip at each end; 1 long edge has a straight area or shallow notch. Cross section ovate. Form ellipsoid, laterally compressed, asymmetrical, often appearing bent to one side. The scutellum is prominent on the dorsal face; it extends about 2/3 the length of the caryopsis.

Surface with scattered shallow pits; faintly striate when seen at 10× magnification. Somewhat shiny at 10×. Translucent orange-brown with darker brown mottling; the rim of the scutellum is black.

Length 0.9–1.7 mm; width (ventral face to dorsal face) 0.7–1.3 mm; thickness 0.5–0.7 mm.
Waste places; sandy or rocky dry soil; E 2/3 GP.

Glossary

Accessory fruit   A fruit derived from flower parts in addition to the ovary.

Accumbent   The embryo bent, with the radicle pressed against the edges of the 2 cotyledons.

Achene   A dry, indehiscent, 1-seeded fruit.

Alveolate   Surface with shallow rounded indentations with narrow ridges between them, having a low relief honeycomblike appearance.

Antrorse   Oriented forward or upward.

Apex   The upper end, or the end opposite the hilum (the attachment point of a fruit) regardless of actual orientation.

Attachment point   The area at the base of a fruit that was attached to the receptacle of the flower. (The attachment point is not a true hilum.)

Awn   A conspicuous terminal bristle.

Base   The lower end, or the attachment end regardless of actual orientation.

Beak   A long, firm tip. In describing the achenes of the Asteraceae, *beak* is used to mean the remnant of the style base that is typically present at the apex. In most achenes of the Asteraceae this beak is small, either blunt or pointed, but in achenes of the Lactuceae the beak is long and blunt.

Bract   A modified reduced leaf in an inflorescence.

Bur   An involucral structure that resembles a fruit but is derived from an inflorescence; it is firm and indehiscent and completely encloses one or more fruits.

Caruncle   A fleshy outgrowth of the funiculus; often conspicuous in seeds of the Euphorbiaceae.

Caryopsis   A grain, the fruit of the Poaceae; in a caryopsis the seed coat is fused to the pericarp.

Chalaza   Area of the ovule where the integuments and nucellus join; the chalaza may be noticeable as a raised or depressed area on the seed surface.

Collar   A thin, raised circular band.

Colliculate   Surface closely covered with small rounded elevations in low relief.

Coma	A tuft of fine hairs.
Cross-corrugate	A distinct pattern of crosswise wrinkles.
Dorsal	The back or under side; the dorsal surface of a leaf, petal, or fruit is the side away from the stem or central axis. Dorsal is the opposite of ventral.
Embryo	The rudimentary immature plant contained in the seed.
Filiform	Slender, threadlike.
Floret	A small flower in a dense inflorescence; in the Poaceae, some of the floret parts may be persistent on the fruit (caryopsis), sometimes completely enclosing it. In the Asteraceae some parts of the floret, e.g., the pappus, may persist on the achene.
Fruit	A mature ovary.
Funiculus	The stalk attaching the ovule to the ovary wall.
Glumes	Sterile bracts subtending a spikelet (Poaceae).
Hilum	The scar on the seed surface where the immature seed (i.e., the ovule) was attached to the funiculus.
Incumbent	The embryo bent, with the radicle pressed against the broad side of the cotyledons.
Integuments	One or two outer layers of the ovule that develop into the seed coat.
Involucre	A cluster of bracts subtending an inflorescence.
Lacerate	An irregular margin, appearing torn.
Lemma	The outer of 2 bracts subtending a grass floret.
Mericarp	A 1-seeded unit of a schizocarp.
Nut	A dry, hard, indehiscent 1-seeded fruit.
Nutlet	A small nut.
Ovary	The ovule-bearing portion of the pistil. When mature, the ovary is termed a fruit.
Ovule	A structure within the ovary that encloses the female gametophyte. It is composed of the nucellus, integuments, funiculus, and embryo sac. A mature ovule is termed a seed.
Palea	The inner of 2 bracts subtending a grass floret.
Papillae	Small, rounded surface projections in moderate relief.
Pappus	In fruits (achenes) of the Asteraceae, an apical crown of scales, teeth, bristles, hairs, or awns; it is a modified calyx.
Perianth	The petals and sepals of a flower collectively.
Pericarp	The fruit wall, derived from the ovary wall.
Plumose	Featherlike, with fine hairs.
Processes	Fine projections or extensions of any kind.
Punctate	Minutely pitted.
Rachilla	The axis of a grass spikelet.
Rachis	The main axis of a grass inflorescence.
Raphe	A ridge or line on the seed surface formed by a part of the funiculus

	that is persistent and fused to the seed coat; typical of seeds of the Euphorbiaceae.
Reticulate	A distinct netlike or honeycomblike surface texture, with the network in sharper relief than that of an alveolate surface.
Retrorse	Oriented backward or downward.
Ribs	Lengthwise straight ridges.
Rugose	Surface with an overall pattern of wrinkles.
Scabrous	Rough or harsh surface texture.
Scalariform	A low relief ladderlike pattern on the surface.
Schizocarp	A dry, dehiscent fruit that separates into 1-seeded units (mericarps) at maturity.
Scutellum	The large cotyledon of a grass embryo. The scutellum is often visible on the surface of the caryopsis as a depressed area near the base.
Seed	A mature ovule. A seed is composed of the embryo and the seed coat and may also include a nutrient storing tissue, either endosperm or perisperm.
Seed coat	Outer covering of the seed, derived from the integument(s) of the ovule.
Seed stalk	Funiculus, the stalk attaching the ovule to the ovary wall.
Spikelet	In the Poaceae, a unit of an inflorescence, composed of 1 to many florets.
Striations	Very fine grooves or scratches.
Strigose	With stiff, straight, appressed hairs that are swollen at the base.
Style	The elongated portion of the pistil, between the ovary and the pollen-receptive area, the stigma.
Stylopodium	An enlarged style base, persistent in the fruit (typical of the Apiaceae).
Tubercles	Wartlike surface projections, in marked relief.
Ventral	The top or upper side; the ventral side of a leaf, petal, or fruit faces the stem or central axis. Ventral is the opposite of dorsal.

Illustrated Glossary

Outline and Cross Sectional Shapes

Round (circular)
Symmetrical in all directions.

Elliptical
Elongate; side margins symmetrically curved; widest at or near the midpoint.

Ovate
Elongate; side margins symmetrically curved; widest below the midpoint.

Obovate
Inverted ovate; widest above the midpoint.

Lanceolate
Narrowly ovate

Oblanceolate
Narrowly obovate

Pear shaped

Tear-drop shaped

Reniform
(kidney shaped)

Cordate
(Heart shaped)

Square
Rectangular and equilateral.

Oblong
Rectangular, elongate; side
margins straight and parallel.

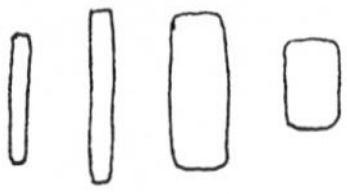

Rhombic
With 4 sides and 4 angles;
widest at or near the midpoint.

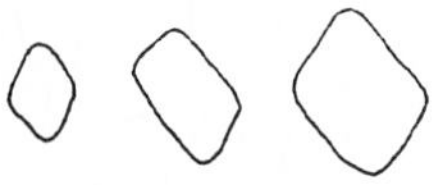

Trapezoidal

Triangular
With 3 sides and 3 angles;
may be symmetrical or not.

Biconvex

Concave-convex

Plano-convex

V-shaped

U-shaped

C-shaped

**Shapes of
Bases and Apices**

Acute
Pointed; the straight
margins form an angle
of 45° to 90°.

Acuminate
Pointed; the straight
margins form an angle
of less than 45°.

Cuspidate
Pointed; the concave
margins form a narrow tip.

Rounded
Smooth curve.

Truncate
As if cut off at right angles
to the longitudinal axis

Three-Dimensional Forms

Globose

Reniform

Ellipsoid

Conical

Ovoid

Chip-like

Terete
(cylindrical)

Discoid

Clavate

Hemispherical

Sector
(of a globose form)

Sector
(of an elongate form)

Surface Textures   **Granular**

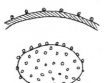

Punctate

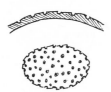

Papillate

Alveolate

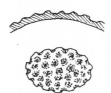

Tuberculate or warty

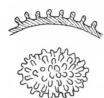

Reticulate

Colliculate

Finely textured

Ridged or wrinkled

Ribbed

Index

The numbers in this index are species numbers, not page numbers.